Praise for Tom Phillips's *Humans*

"Thoroughly informative. Thoroughly entertaining. Thoroughly demoralizing. In a fun kind of way."

—Robert Sapolsky, *New York Times* bestselling author of *Behave: The Biology of Humans at Our Best and Worst*

"*Humans* is a thoroughly entertaining account of unintended consequences, of arrogance and ignorance, of human follies and foibles from ancient times to the present. It seems history has taught us nothing—we are doomed to keep suffering the antics of both well-intended and ill-intended fools. As I was reading I wondered how I could be so disheartened and yet at the same time be laughing out loud."

—Penny Le Couteur, author of *Napoleon's Buttons: How 17 Molecules Changed History*

"Tom Phillips has proven beyond a doubt that humans are goddamn lucky to be here and are doing nearly nothing to remain relevant and viable as a species—except, that is, for writing witty, entertaining, and slightly distressing-but-ultimately-endearing books about the same. And if you care to avoid orbiting the earth in a space-garbage prison of your fellow humans' design, you should probably read it."

—Sarah Knight, *New York Times* bestselling author of *Get Your Sh*t Together*

"*Humans* is Tom Phillips' timely, irreverent gallop through thousands of years of human stupidity. Every time you begin to find our foolishness bizarrely comforting, Phillips adds another kick in the ribs. Beneath all this book's laughter is a serious question: Where does so much serial stupidity take us?"

—Nicholas Griffin, author of *Ping-Pong Diplomacy: The Secret History Behind the Game That Changed the World*

"A laugh-along, worst-hits album for humanity. With the delicate touch of a scholar and the laugh-out-loud chops of a comedian, Tom Phillips shows how our species has been messing things up ever since we evolved from apes and came down from the trees some four million years ago."

—Steve Brusatte, University of Edinburgh paleontologist and *New York Times* bestselling author of *The Rise and Fall of the Dinosaurs*

HUMANS:

A BRIEF HISTORY OF HOW WE F[illegible]ED IT ALL UP

HUMANS: A BRIEF HISTORY OF HOW WE F███ED IT ALL UP

Tom Phillips

HANOVER
SQUARE
PRESS

ISBN-13: 978-1-335-93663-9

Humans: A Brief History of How We F*cked It All Up

Originally published in Great Britain by Wildfire in 2018
This edition published by Hanover Square Press in 2019

Library of Congress Cataloging-in-Publication Data has been applied for.

HanoverSqPress.com
BookClubbish.com

Printed in U.S.A.

Given the subject matter,
dedicating this book to my family
could be badly misinterpreted.

So instead, I dedicate this to anybody
who has ever fucked up really badly.
You are not alone.

Contents

Timeline of History

3200000 BCE	Lucy falls out of a tree and dies. Humanity will repeat this pattern many times over the following 3.2 million years.
70000 BCE	Modern humans migrate out of Africa, ruining everything for everyone else.
70000 BCE – 40000 BCE	Really bad period for Neanderthals.
12000 BCE	Humanity invents war! YAY HUMANITY! GO TEAM!
11000 BCE	Agriculture is invented, which may also have been an awful mistake tbh.
3000 BCE	The Sumerians and Egyptians invent the idea of "absolute dynastic monarchy." Thanks for that, Egypt and Sumer!
2334 BCE	Sargon of Akkad goes one further and invents the idea of "empires." Thanks for that, Sargon!
222 BCE	Qin Shi Huang unites China, searches for elixir of life, dies.
216 BCE	Battle of Cannae. Romans experiment with having two leaders, with opposing strategies. Goes roughly as well as you'd guess.
27 BCE	In excellent news for fans of dictatorship, the Roman Republic becomes the Roman Empire.
26 BCE – 892 AD	Not much happens. Pretty quiet time, historically speaking.
1004	First contact between Europeans and Americans; ends in lots of murder.
1217	Ala ad-Din Muhammad II makes the worst decision in history: making an enemy of Genghis Khan.

1492	Christopher Columbus fails to discover new route to Asia, crashes into America instead. Honestly this is the point where everything starts going really wrong.
1519	In history's most ill-advised hospitality, Moctezuma invites Cortés in as a guest.
1617 – 1648	Ottoman Empire endures a run of mostly terrible leaders (two of them are called "the Mad," which is a bad sign).
1698	Scotland tries to establish an empire in Panama. This doesn't go well, leaving hundreds dead and the country almost bankrupt.
1788	Austrian army manages to defeat itself at the Battle of Karansebes.
1812	Napoleon tries to invade Russia. This turns out to be a terrible idea.
1859	Thomas Austin introduces 24 rabbits into Australia. This doesn't end well.
1885	King Leopold II is given the Congo for charitable purposes. His purposes are...not charitable.
1890	Shakespeare fan Eugene Schieffelin introduces 60 starlings to New York, whimsically. They become a major, non-whimsical pest.
1914	World goes to war. It's awful.
1917	In a well thought-through plan, Germany helps Lenin get back to Russia.
1923	The first leaded gasoline, developed by Thomas Midgley Jr., goes on sale. Several generations get lead poisoning.
1928	Not one to rest on his laurels, Thomas Midgley Jr. develops Freon. Which is bad news for the ozone layer.

1929	It is predicted that the economy is doing very well. Global financial crisis begins a few days later.
1933	The first dust storms of the American dust bowl begin.
1933	The very cunning German politician Franz von Papen does a deal wth Hitler in an attempt to regain power. Yeah, that doesn't work out great.
1939	World goes to war again. Even more awful this time.
1941	Hitler tries to invade Russia. Luckily, this turns out to still be a terrible idea.
1945	Robert Oppenheimer predicts that nuclear weapons will end war. Results so far are mixed.
1958	Mao's Four Pests campaign begins, leading to the killing of 1 billion sparrows.
1959	Chinese famine begins, caused in part by sudden lack of sparrows.
1960	Soviets divert rivers from Aral Sea. In shock news, the Aral Sea dries up.
1961	USA hilariously fails to invade Cuba at the Bay of Pigs.
1981	The Californian town of Sunol elects a dog called Bosco Ramos as mayor. This is the only good thing to happen on this timeline.
2007	It is predicted that the economy is doing very well. Global financial crisis begins a few days later.
2018	In April 2018, atmospheric carbon dioxide passes 410 parts per million for the first time in 3.2 million years. Hey, Lucy!
2019	This book is published. A new age of enlightenment dawns.

PROLOGUE

THE DAWN

OF

FUCK-UPS

A long, long time ago, as the sun rose across the great river valleys and plains of Ethiopia, a young ape was lounging in a tree.

We can't know what she was thinking or doing that day. Probably she was pondering finding something to eat, or finding a mate, or perhaps checking out the next tree over to see if it was a better tree. She certainly didn't know that the events of that day would make her the most famous member of her species ever—even if you could somehow tell her, the concept of fame wouldn't make any sense to her. She also didn't know that she was in Ethiopia, because this was millions of years before anybody had the bright idea of drawing lines on a map and giving the shapes names that we could have wars about.

She and her kin were slightly different from the other apes that lived at the same time: there was something unusual about their hips and legs that let them move in a novel way. These apes were beginning their descent from the trees, and starting to walk upright across the savannahs: the initial change that, in time, would lead to you and me and every other person on this planet. The ape didn't know it, but she was living near the beginning of one of the most remarkable stories ever. This was the dawn of the great human journey.

Then she fell out of the tree and died.

Roughly 3.2 million years later, a different group of apes—some of them now in possession of PhDs—would dig up her fossilized bones. Because this was the 1970s, and they were listening to a popular song by a group of extremely high Liverpudlians at the time, they decided to call her Lucy. She was a brand-new species—what we now call an *Australopithecus afarensis*—and she was hailed as the "missing link" between humans and apes. Lucy's discovery would captivate the world: she became a household name, her skeleton would be taken on a multiyear tour of the USA and she's now the star attraction in the National Museum of Ethiopia, in Addis Ababa.

And yet the only reason we know about her is because, bluntly, she fucked up. Which in retrospect set a pretty clear template for how things were going to play out from that point onward.

This is a book about humans, and our remarkable capacity for fucking things up. About why for every accomplishment that makes you proud to be human (art, science, tacos), there's always something else that makes you shake your head in bafflement and despair (war, pollution, Taco Bell).

It's quite likely that—regardless of your personal opinions or political persuasion—at some point in recent times you've looked around at the state of the world and thought to yourself: Oh, shit, what have we done?

This book is here to provide a tiny, hollow grain of comfort: don't worry, we've *always* been like this. And hey, we're still here!

(Granted, at the time of writing this, there's a broad awareness that the only thing that stands between us and imminent fiery nuclear annihilation is the whim of one petulant man-child or another, so who knows? I'm going to work on the assumption that you are in fact reading this book in the normal way, rather than planning to burn it for warmth as you shelter in the rubble, in which case I guess we made it to May 2019 at the very least.)

There are lots of books about humanity's finest achievements—the great leaders, the genius inventors, the indomitable human spirit. There are also lots of books about mistakes we've made: both individual screw-ups and society-wide errors. But there aren't quite so many about how we manage to get things profoundly, catastrophically wrong over and over again.

In one of those ironies that the universe seems to really enjoy, the reasons we cock it up on such a vast scale are often the exact same things that set us apart from our fellow animals and allow us to achieve greatness. Humans see patterns in the world, we can communicate this to other humans and we have the capacity to imagine futures that don't yet exist: how if we just changed *this thing*, then *that thing* would happen, and the world would be a slightly better place.

The only trouble is...well, we're not terribly good at any of those things. Any honest assessment of humanity's previous performance on those fronts reads like a particularly brutal annual review from a boss who hates you. We imagine patterns where they don't exist. Our communication skills are, uh, sometimes lacking. And we have an extraordinarily poor track record of failing to realize that changing *this thing* will also lead to *the other thing*, and *that even worse thing*, and *oh God no now this thing is happening how do we stop it*.

No matter how high humanity rises, no matter how many challenges we conquer, catastrophe is always lurking just around the corner. To pick a historical example: one moment you are Sigurd the Mighty (a ninth-century Norse Earl of Orkney), riding home in triumph from battle with the head of your slain enemy, Máel Brigte the Bucktoothed, dangling from your saddle.

The next moment, you are...well, you're Sigurd the Mighty a couple of days later, dying from an infection caused when the protruding bucktooth of Máel Brigte the Bucktoothed's disembodied head grazed your leg while you were riding home in triumph.

That's right: Sigurd the Mighty holds the dubious distinction in military history of being killed by an enemy he'd already decapitated several hours earlier. Which teaches us important lessons about (a) hubris, and (b) the importance of choosing enemies who have high-quality dental care. It's hubris and its subsequent downfalls that will be the major focus of this book. Fans of historical dentistry standards, by contrast, may be sadly disappointed.

(It's also worth noting that Sigurd the Mighty and Máel Brigte the Bucktoothed were only fighting because Sigurd had challenged Máel Brigte to a "forty soldiers on each side" battle. Máel Brigte agreed to this, whereupon Sigurd turned up with eighty soldiers. As such, there is possibly also a lesson in Sigurd's story about the importance of not being a complete dick, which funnily enough is also a theme that recurs throughout the book.)

Sigurd is just one of the many unfortunates who history remembers more for their losses than their wins. Over the next ten chapters, we'll take a tour of the entire sweep of human history, and its catalog of cock-ups. A gentle warning: if you're not really into schadenfreude, now might be a good time to stop reading.

The story of human progress starts with our capacity for thinking and creativity. That's what sets humans apart from other animals—but it's also what leads us to make complete tits of ourselves on a regular basis.

In the first chapter of this book, "Why Your Brain Is an Idiot," we'll look at how our ancestors thought differently—and then see how our attempts to make sense of the world end up with our minds playing tricks on us, letting us down and leading us into making all our terrible, terrible decisions.

Then in the second chapter, "Nice Environment You've Got Here," we'll follow humanity to the dawn of agriculture, as we started to shape the world around us, and see how we regularly make a complete mess of the places we live, tracing our unfailing ability to not really think through the answer to the question: Hey, what's the worst that could happen if we divert this river?

After that, we'll check in on our consistently ham-fisted attempts to control nature, in "Life, Uh, Finds a Way"—where we get to see, among other things, how Chairman Mao and a whimsical Shakespeare enthusiast managed to cause mirror-image catastrophes by radically underestimating birds.

As humanity's earliest societies developed and grew more complex, it became apparent that we were going to need someone in charge of making decisions. In the fourth chapter, "Follow the Leader," we'll look at a selection of the absolute worst unelected people to have ever had that job; in Chapter 5, "People Power," we'll check in on democracy to see if that does any better.

For all that we manage to shape the world around us, humanity's true potential for looking like complete idiots was not fully realized until we traveled the world and different civilizations started meeting each other. That's when we got to really let our hair down and get things profoundly, catastrophically wrong.

In Chapter 6, "War. Huh. What Is It Good For?" we'll see how humans have a very long history of getting into pointless fights, and examine some of the stupidest things to have happened as a result—including the army that managed to lose a battle their opponents didn't even show up for, and how to mess up your perfectly coordinated attack plans by forgetting that time zones exist.

We'll head out into the unknown with the heroic figures of the Age of Discovery in Chapter 7, "Super Happy Fun Colonialism Party," in which we will discover (spoiler alert) that colonialism was terrible.

Chapter 8, "A Dummies' and/or Current Presidents' Guide to Diplomacy," will teach us important lessons about how to gracefully handle contacts between different cultures, including how the shah of the Khwarezmian Empire made possibly the single worst political decision in history. (It involved setting beards on fire.)

In recent centuries, scientific and technological advances have ushered in an era of unprecedented innovation, rapid change and exciting new ways for humanity to fail. That's the focus of Chapter 9, "The Shite Heat of Technology," where we'll see how science doesn't always get things right—including the mysterious radiation that only French people could see, and the man who made not one but two of the twentieth century's most catastrophic mistakes.

Change now happens so quickly that the modern world can be a confusing place; in Chapter 10, "A Brief History of Not Seeing Things Coming," we'll look back at exactly how frequently we've failed to predict the awful new things that are about to happen to us.

And finally, in "Fucking Up the Future," we'll take an educated guess at what the next few centuries of human foolishness will look like, and conclude that it probably means becoming trapped in a space prison we've made for ourselves out of our own garbage.

This is a book about history, and about getting things wrong. So naturally, it's worth pointing out that we often get history very, very wrong.

The problem is that history is slippery: nobody bothered to write down the vast majority of stuff that happened in it, and lots of the people who did write stuff down might have been mistaken, or mad, or lying, or extremely racist (and frequently a combination of all those things). We know about Sigurd the Mighty because his story appears in two documents, the sagas of Heimskringla and Orkneyinga. But how do we know if they're accurate? Can we be entirely sure that this wasn't just some sort of extremely funny Old Norse in-joke that we don't get?

We can't. Not really, despite the amazing work done by historians and archaeologists and experts in a dozen other fields. The number of things that we know for certain is pretty tiny compared to the number of things that we know we don't know. The number of things that we don't even know we don't know is probably far bigger still, but unfortunately we don't know for sure.

What I'm saying is: the chance of this book about fuck-ups not including any fuck-ups in it is, frankly, minimal. I'll try to make it clear where there's uncertainty: which are the bits we're pretty sure about, and which are the bits where the best we can do is an educated guess. I've tried to avoid any "too good to be true" stories, the apocryphal tales and pithy historical anecdotes that seem to grow with each retelling. I hope I don't get it wrong.

Which brings us back to Lucy, falling out of her tree 3.2 million years ago. How do we know she fell out of that tree? Well, in 2016, a group of researchers from the USA and Ethiopia published a paper in *Nature*, the world's leading scientific journal. They CT-scanned Lucy's fossilized bones, creating 3-D computer maps of them to reconstruct her skeleton. They found that the fractures in her bones were the kind that happen to living bones, and that these fractures never healed: suggesting that she was alive when they broke but died soon after. They consulted numerous orthopedic surgeons, who all said the same thing: this is the pattern of broken bones that you see in a patient who has fallen from a height. The way her arm is fractured suggests that she reached out to try to break her fall. From geological studies, they knew the area she lived in was flat woodland, near a stream: no cliffs or outcrops for her to fall off. The conclusion? Lucy fell out of a tree.

It's a remarkable piece of work, and one that was well received by many other experts in the field. The only problem is that a few other experts—including Donald Johanson, the man who discovered Lucy in the first place—weren't convinced. They effectively said: "Nah, mate, the reason her bones are broken is because that's what happens to bones when they're buried in the ground for 3.2 million years." (I'm paraphrasing a bit here.)

So...did Lucy fall out of a tree? Maybe. Probably, even. In many ways, that's the point of this book: we have this incredible feat of scientific deduction, and it *still* might be wrong. You can be a world leader in your field, doing the best work of your career, a groundbreaking study published in the world's most prestigious journal that weaves together jaw-dropping advances in the fields of paleontology and physics, computing and medicine, forensics and geology, to give us an unprecedented window into a time millions of years ago...and you still run the risk that someone will come along and go: "Hahahaha, nope."

Just when you think you've got it all sorted out, that's when the ever-looming specter of fuck-ups will strike.

Remember Sigurd the Mighty.

1

WHY YOUR BRAIN IS AN IDIOT

It was about 70,000 years ago that human beings first started to really ruin things for everybody.

That's when our ancestors began to migrate out of Africa and spread across the globe—first into Asia, and a while later into Europe. The reason this made a lot of people rather unhappy is that back then our species, *Homo sapiens*, weren't the only humans on the planet—far from it. Exactly how many other species of humans were knocking around at that point is a matter of some debate. The business of taking fragmentary skeletons or fragmentary DNA and trying to work out exactly what counts as a separate species, or subspecies, or just a slightly weird version of the same species, is a tricky one. (It's also an ideal way to start an argument should you ever find yourself among a group of paleoanthropologists with some time to kill.) But however you classify them, there were at least a couple of other types of humans on the planet back then, of which the most famous is *Homo neanderthalensis*—or, as they're better known, the Neanderthals. The result of previous human migrations from Africa, they'd been living across much of Europe and large parts of Asia for over 100,000 years. They basically had quite a good thing going.

Unfortunately for them, just a few tens of thousands of years after our ancestors rocked up on the scene—the blink of an eye in evolutionary terms—the Neanderthals and all our other relatives were gone from the face of the earth. In a pattern that would quickly establish itself throughout human history, as soon as we arrive, there goes the neighborhood. Within a few thousand years of modern humans moving into an area, the Neanderthals start to vanish from the fossil record, leaving behind only a few ghostly genes that still haunt our DNA. (There was clearly a bit of interbreeding between the Neanderthals and the interlopers who were replacing them; if you're of European or Asian descent, for example, there's a good chance that somewhere between 1 and 4 percent of your DNA is Neanderthal in origin.)

Exactly why and how we survived while our cousins got the fast train to Extinctionville is another subject of debate. In fact, lots of the most likely explanations are themes that will keep cropping up again and again in this book. We might have accidentally wiped out the Neanderthals by bringing diseases with us as we migrated that they didn't have any resistance to. (A large part of the history of humanity is really just the history of the diseases we manage to pick up on our travels and then give to each other.) We might have got lucky with a fluctuating climate that we were better able to adapt to; the evidence suggests our ancestors lived in bigger social groups, and communicated and traded over a much larger area than the more isolated, stick-in-the-mud Neanderthals, meaning they could draw on more resources when a cold spell hit.

Or maybe we just murdered them, because, hey, that's what we do.

In all likelihood there probably isn't a single neat explanation, because that's not how things normally work. But many of the most plausible explanations have one thing in common—our brains, and how we use them. It's not quite as simple as the idea that "we were smart and they were dumb"; Neanderthals weren't the lumbering numbskulls of popular stereotype. They had brains as big as we do, and were making tools, controlling fire and producing abstract art and jewelry in Europe tens of thousands of years before *Homo sapiens* ever came along and started gentrifying everything. But most of the plausible advantages we had over our Neanderthal cousins relate to our thinking, whether that's in our adaptability, our more advanced tools, our more complex social structures or the ways we communicated within and between groups.

There's something about the way we humans think that marks us out as special. I mean, obviously. It's right there in the name of our species: *Homo sapiens* is Latin for "wise man." (Modesty, let's be honest, has never really been one of our species' defining traits.)

And in fairness to our egos, the human brain is a truly remarkable machine. We can spot patterns in our environment and make educated guesses from those about the way things work, building up a complex mental model of the world that includes more than what we can see with our eyes. Then we can build upon that mental model to make imaginative leaps: we're able to envisage the changes to the world that would improve our situation. We can communicate these ideas to our fellow humans, so that others can make improvements to them that we wouldn't have thought of, turning knowledge and invention into a communal effort that gets passed down the generations.

After that, we can convince others to work collectively in the service of a plan that previously existed only in our imagination, in order to achieve breakthroughs that none of us could have made alone. And then we repeat this many times in a hundred thousand different ways, over and over again, and what were once wild innovations turn into traditions, which spawn new innovations in turn, until eventually you end up with something that you'd call "culture" or "society."

Think of it this way: the first step is noticing that round things roll down hills better than jagged lumpy things. The second is working out that if you use a tool to chip away at something and make it more round, it'll roll better. The third step is showing your friend your new round rolling things, whereupon they come up with the idea of putting four of them together to make a wagon. The fourth step is building a fleet of ceremonial chariots, so that the people may better understand the glory of your benevolent yet merciless rule. And the fifth step is being stuck in a traffic jam on the New Jersey Turnpike listening to talk radio while flipping off some asshole who's put Truck Nutz on the back of his SUV.

(IMPORTANT NOTE IN THE INTERESTS OF PEDANTRY: this is a wildly inaccurate cartoon description of the invention of the wheel. Wheels actually get invented surprisingly late in the scheme of things, well after civilization has been cheerfully muddling along without them for thousands of years. The first wheel in archaeological history, which pops up about 5,500 years ago in Mesopotamia, wasn't even used for transport: it was a potter's wheel. It seems to have been several hundred more years before somebody had the bright idea of turning potters' wheels on their side and using them to roll stuff around, thus beginning the process that would ultimately lead to assholes who put Truck Nutz on their SUVs. Apologies to any wheel scholars who were offended by the previous paragraph, which was intended for illustrative purposes only.)

But while the human brain is remarkable, it is also extremely weird, and prone to going badly wrong at the worst possible moment. We routinely make terrible decisions, believe ridiculous things, ignore evidence that's right in front of our eyes and come up with plans that make absolutely no sense. Our minds are capable of imagining concertos and cities and the theory of relativity into existence, and yet apparently incapable of deciding which type of potato chips we want to buy at the shop without five minutes' painful deliberation.

How has our unique way of thinking allowed us to shape the world to our desires in incredible ways, but also to consistently make absolutely the worst possible choices despite it being very clear what bad ideas they are? In short: How can we put a man on the moon and yet still send THAT text to our ex? It all boils down to the ways that our brains evolved.

The thing is that evolution, as a process, is not smart—but it is at least dumb in a very persistent way. All that matters to evolution is that you survive the thousand possible horrible deaths that lurk at every turn for just long enough to ensure that your genes make it through to the next generation. If you manage that, job done. If not, tough luck. This means that evolution doesn't really do foresight. If a trait gives you an advantage *right now*, it'll be selected for the next generation, regardless of whether or not it's going to end up lumbering your great-great-great-great-great-grandchildren with something that's woefully outdated. Equally, it doesn't give points for prescience—saying, "Oh, this trait is kind of a hindrance now, but it'll come in really useful for my descendants in a million years' time, trust me" cuts absolutely no ice. Evolution gets results not by planning ahead, but rather by simply hurling a ridiculously large number of hungry, horny organisms at a dangerous and unforgiving world and seeing who fails least.

This means that our brains aren't the result of a meticulous design process aimed at creating the best possible thinking machines; instead, they're a loose collection of hacks and bodges and shortcuts that made our distant ancestors 2 percent better at finding food, or 3 percent better at communicating the concept "Oh shit, watch out, it's a lion."

Those mental shortcuts (they're called "heuristics," if you want to get technical) are absolutely necessary for surviving, for interacting with others and for learning from experience: you can't sit down and work out everything you need to do from first principles. If we had to conduct the cognitive equivalent of a large-scale randomized control trial every time we wanted to avoid being shocked by the sun rising in the morning, we'd never have got anywhere as a species. It's a lot more sensible for your brain to go, "Oh yeah, sun rises" after you've seen it happen a few times. Likewise, if Jeff tells you that eating the purple berries from that bush by the lake made him violently ill, it's probably best to just believe him, rather than try it out for yourself.

But this is also where the problems begin. As useful as they are, our mental shortcuts (like all shortcuts) will sometimes lead us down the wrong path. And in a world where the issues we have to deal with are a lot more complex than "Should I eat the purple berries?" they get it wrong a *lot*. To be blunt, much of the time your brain (and my brain, and basically everybody's brain) is a massive idiot.

For a start, there's that ability to spot patterns. The problem here is that our brains are so into spotting patterns that they start seeing them all over the place—even where they don't exist. That's not a huge problem when it just means stuff like pointing at the stars in the night sky and going, "Ooh, look, it's a fox chasing a llama." But once the imaginary pattern you're seeing is something like "most crimes are committed by one particular ethnic group," it's...well, it's a really big problem.

There are a bunch of terms for this kind of faulty pattern-spotting—things like "illusory correlation" and the "clustering illusion." During World War II, many people in London became convinced that German V-1 and V-2 missiles (an already pretty terrifying new technology) were falling on the city in targeted clusters—leading Londoners to seek shelter in supposedly safer parts of the city, or suspect that certain seemingly untouched neighborhoods housed German spies. This was concerning enough that the British government got a statistician named R. D. Clarke to check whether it was true.

His conclusion? The "clusters" were no more than our minds playing tricks on us, the illusory ghosts of pattern-matching. The Germans hadn't made a dramatic breakthrough in guided missile technology, after all, and Clerkenwell was not a hotbed of Wehrmacht secret agents; the doodlebugs were just being lobbed in the general direction of the city entirely at random. People only saw patterns because that's what our brains do.

Even skilled professionals can fall victim to these types of illusions. For example, plenty of medical workers will tell you with certainty that a full moon invariably leads to a bad night in the ER ward—a surge of patients, bizarre injuries and psychotic behavior. The only trouble is that studies have looked at this, and as far as they can tell, it's just not true: there's no link between the phases of the moon and how busy emergency rooms get. And yet a bunch of talented, experienced professionals will swear blind that there is a connection.

Why? Well, the belief doesn't come from nowhere. The idea that the moon makes people go weird is one that's been around for centuries. It's literally where the word *lunacy* comes from; it's why we have werewolf mythology. (It may also be related to the supposed correlation between the phases of the moon and women's menstrual cycles.) And the thing is, it actually might have been sort of true at one time! Before the invention of artificial lighting—street lighting especially—the light of the moon had a much greater effect on people's lives. One theory suggests that homeless people sleeping outdoors would have been kept awake by the full moon, with sleeplessness exacerbating any mental health problems they had. (Because I like theories that involve beer, I'd also float an alternative suggestion: people probably got way more drunk on evenings when they knew they could see their way home and so were less worried about getting lost, or robbed, or tripping over and dying in a ditch.)

Wherever it comes from, it's an idea that's been fixed in culture for a long time. And once you've been told about the idea that the full moon means crazytime, you're much more likely to remember all the times that it *did* happen—and forget the times it didn't. Without meaning to, your brain has created a pattern out of randomness.

Again, this is because of those mental shortcuts our brains use. Two of the main shortcuts are the "anchoring heuristic" and the "availability heuristic," and they both cause us no end of bother.

Anchoring means that when you make up your mind about something, especially if you don't have much to go on, you're disproportionately influenced by the first piece of information you hear. For example, imagine you're asked to estimate how much something costs, in a situation where you're unlikely to have the knowledge to make a fully informed judgment—say, a house you're shown a picture of. (Note for millennials: houses are those big things made of bricks you'll never be able to buy.) Without anything else to go on, you might just look at the picture, see roughly how fancy it looks and make a wild stab in the dark. But your guess can be dramatically skewed if you're given a suggested figure to begin with—for example, in the form of a preceding question such as "Do you think this house is worth more or less than $400,000?" Now, it's important to realize that question hasn't actually given you any useful information at all (it's not like, say, being told what other houses in the area have recently sold for). And yet people who get prompted with a figure of $600,000 will end up estimating the house's value much higher on average than people who are prompted with $200,000. Even though the preceding question isn't informative at all, it still affects your judgment, because you've been given an "anchor"—your brain seizes on it as a starting point for making its guess, and adjusts from there.

We do this to an almost ridiculous degree: the piece of information we use as an anchor can be as explicitly unhelpful as a randomly generated number, and our brains will still latch on to it and skew our decisions toward it. This can get frankly worrying; in his book *Thinking, Fast and Slow*, Daniel Kahneman gives the example of a 2006 experiment on a group of highly experienced German judges. They were shown details of a court case in which a woman was found guilty of shoplifting. They were then asked to roll a pair of dice, which (unknown to them) were weighted to only ever produce a total of 3 or 9. Then they were asked if the woman should be sentenced to more or fewer months than the figure produced by the dice, before finally being asked to provide their best recommendation for how long her sentence should be.

You can pretty much guess the result: the judges who rolled the higher figure on the dice sentenced her to much longer in prison than the ones who rolled low. On average, the roll of the dice would have seen the woman spend an extra three months in jail. This is not comforting.

Availability, meanwhile, means that you make judgment calls on the basis of whatever information comes to mind easiest, rather than deeply considering all the possible information that might be available to you. And that means we're hugely biased toward basing our worldview on stuff that's happened most recently, or things that are particularly dramatic and memorable, while all the old, mundane stuff that's probably a more accurate representation of everyday reality just sort of...fades away.

It's why sensational news stories about horrible crimes make us think that crime levels are higher than they are, while dry stories about falling crime statistics don't have anywhere near as much impact in the opposite direction. It's one reason why many people are more scared of plane crashes (rare, dramatic) than they are of car crashes (more common and as such a bit less exciting). And it's why terrorism can produce instant knee-jerk responses from the public and politicians alike, while far more deadly but also more humdrum threats to life get brushed aside. More people were killed by lawn mowers than by terrorism in the USA in the decade between 2007 and 2017, but at the time of writing, the US government has yet to launch a War on Lawn Mowers. (Although, let's be honest, given recent events you wouldn't rule it out.)

Working together, the anchoring heuristic and the availability heuristic are both really useful for making snap judgments in moments of crisis, or making all those small, everyday decisions that don't have much impact. But if you want to make a more informed decision that takes into account all the complexity of the modern world, they can be a bit of a nightmare. Your brain will keep trying to slide back to its evidential comfort zone of whatever you heard first, or whatever comes to mind most quickly.

They're also part of the reason why we're terrible at judging risk and correctly predicting which of the many options available to us is the one least likely to lead to catastrophe. We actually have two separate systems in our minds that help us judge the danger of something: the quick, instinctive one and a slow, considered one. The problems start when these conflict. One part of your brain is quietly saying, "I've analyzed all the evidence and it appears that Option 1 is the riskiest alternative," while another part of your brain is shouting, "Yes, but Option 2 SEEMS scary."

Sure, you might think, but luckily we're smarter than that. We can force our brains out of that comfort zone, can't we? We can ignore the instinctive voice and amplify the considered voice, and so objectively consider our situation, right? Unfortunately, that doesn't take confirmation bias into account.

Before I began researching this book, I thought that confirmation bias was a major problem, and everything I've read since then convinces me that I was right. Which is exactly the problem: our brains *hate* finding out that they're wrong. Confirmation bias is our annoying habit of zeroing in like a laser-guided missile on any scrap of evidence that supports what we already believe, and blithely ignoring the possibly much, much larger piles of evidence that suggest we might have been completely misguided. At its mildest, this helps explain why we prefer to get our news from an outlet that broadly agrees with our political views. In a more extreme instance, it's why you can't argue a conspiracy theorist out of their beliefs, because we cherry-pick the events that back up our version of reality and discard the ones that don't.

Again, this is quite helpful in some ways: the world is complex and messy and doesn't reveal its rules to us in nice, simple PowerPoint presentations with easy-to-read bullet points. Coming up with any kind of mental model of the world means discarding useless information and focusing on the right clues. It's just that working out what information is the stuff worth paying attention to is a bit of a cognitive crapshoot.

It gets worse, though. Our brain's resistance to the idea that it might have screwed up goes deeper. You'd think that once we'd made a decision, put it into action and *actually seen it start to go horribly wrong*, we would then at least become a bit better at changing our minds. Hahaha, no. There's a thing called "choice-supportive bias," which basically means that once we've committed to a course of action, we cling on to the idea that it was the right choice like a drowning sailor clinging to a plank. We even replay our memories of how and why we made that choice in an attempt to back ourselves up. In its mild form, this is why you end up hobbling around in agony after buying a new pair of shoes, insisting to everybody that "they make me look POWERFUL yet ALLURING." In a stronger form, it is why government ministers continue to insist that the negotiations are going very well and a lot of progress has been made even as it becomes increasingly apparent that everything is going quite profoundly to shit. The choice has been made, so it must have been the right one, because we made it.

There's even some evidence that, in certain circumstances, the very act of telling people they're wrong—even if you patiently show them the evidence that clearly demonstrates why this is the case—can actually make them believe the wrong thing *more*. Faced with what they perceive as opposition, they double down and entrench their beliefs even more strongly. This is why arguing with your racist uncle on Facebook, or deciding to go into journalism, may be an ultimately doomed venture that will only leave you despondent and make everybody else very angry with you.

None of this means that people can never make sensible and well-informed decisions: very obviously they can. I mean, you're reading this book, after all. Congratulations, you excellent choice-maker! It's just that our brains often put a remarkably large number of obstacles in the way, all the time thinking they're being helpful.

Of course, if we're bad at making decisions on our own, it can get even worse when we make decisions along with other people. We're a social animal, and we reeeaaalllly don't like the feeling of being the odd one out in a group. Which is why we frequently go against all our better instincts in an effort to fit in.

That's why we get groupthink—when the dominant idea in a group overwhelms all the others, dissent being dismissed or never voiced thanks to the social pressure to not be the one saying, "Uh, I'm not sure this is the greatest idea?" It's also why we jump on bandwagons with wild abandon: the very act of seeing other people do or believe a thing increases our desire to match them, to be part of the crowd. When your mum asked you as a kid, "Oh, and if the other kids jumped off a bridge, would you do that, too?" the honest answer was, "Actually, there's a pretty good chance, yeah."

And finally, there's the fact that—bluntly—we think we're pretty great when we are not, in fact, all that. Call it hubris, call it arrogance, call it being a bit of a pillock: research shows that we wildly overestimate our own competence. If you ask a group of students to predict how high up the class they'll finish, the overwhelming majority will put themselves in the top 20 percent. Hardly any will say, "Oh yeah, I'm probably below average." (The most common answer is actually outside the top 10 percent, but inside the top 20 percent, like a boastful version of ordering the second-cheapest glass of wine.)

There's a well-known cognitive problem called the Dunning–Kruger effect, and beyond sounding like an excellent name for a seventies prog rock band, it may be the patron saint of this book. First described by the psychologists David Dunning and Justin Kruger in their paper "Unskilled and Unaware of It: How Difficulties in Recognizing One's Own Incompetence Lead to Inflated Self-Assessments," it provides evidence for something that every one of us can recognize from our own lives. People who are actually good at any particular skill tend to be modest about their own abilities; meanwhile, people with no skills or talent in the field wildly overestimate their own competence at it. We literally don't know enough about our own failings to recognize how bad we are at them. And so we blunder on, overconfident and blissfully optimistic about whatever it is we're about to get horribly, horribly wrong. (As the rest of this book will show, of all the mistakes our brains make, "confidence" and "optimism" may well be the most dangerous.)

All of these cognitive failures, piled one on top of the other in the form of society, lead us to make the same types of mistakes over and over again. Below are just a few of them: think of this like a spotter's guide for the rest of the book.

For starters, our desire to understand the world and discern patterns in it means that we spend quite a lot of our time convincing ourselves that the world works a certain way when in fact it absolutely doesn't work like that. This can encompass everything from small personal superstitions to completely inaccurate scientific theories, and explains why we fall for propaganda and "fake news" so readily. The real fun starts when somebody manages to convince large numbers of other people that their pet theory about how the world works is true, which gives you religion and ideology and all those other Big Ideas that have proved so entertaining over the course of human history.

Humans are also very bad at risk assessment and planning ahead. That's partly because the art of prediction is notoriously difficult, especially if you're trying to make predictions about a highly complex system, like the weather or financial markets or human society. But it's also because once we've imagined a possible future that pleases us in some way (often because it fits with our preexisting beliefs), we'll cheerfully ignore any contrary evidence and refuse to listen to anybody who suggests we might be wrong.

One of the strongest motivators for this kind of wishful-thinking approach to planning is, of course, greed. The prospect of quick riches is one that's guaranteed to make people lose all sense—it turns out we're very bad at doing a cost–benefit analysis when the lure of the benefit is too strong. Not only will humans cross oceans and climb mountains for the (often fanciful) promise of wealth, we'll also happily cast aside any notion of morality or propriety as we do it.

Greed and selfishness also play into another common mistake: that of us collectively ruining things for everybody because we each wanted to get an advantage for ourselves. In social science, this category of screw-ups goes by names like the "social trap" or the "tragedy of the commons," which is basically when a group of people all do things that on their own would be absolutely fine in the short term, but when lots of people do them together, it all goes horribly wrong in the long term. Often this means destroying a shared resource because we exploit it too much: for example, fishing an area of water so much the fish stocks can't replenish themselves. There's also a related concept in economics known as a "negative externality"—basically a transaction where both parties do well out of it, but there's a cost that's paid elsewhere, by someone who wasn't even part of the transaction. Pollution is the classic example of that; if you buy something from a factory, that's a win for both you and the manufacturer, but it might be a loss for the people who live downstream of the toxic waste the factory pours out.

This group of related mistakes are behind a really large proportion of human fuck-ups—in systems from capitalism to co-operatives, and from issues that can be as vast as climate change or as small as splitting the bill in a restaurant. We know that it's a bad idea for everyone to underplay how much they owe, but if everyone else is doing it, we don't want to be the ones to lose out by not doing it. And so we shrug, and say, "Not my problem, mate."

Another one of our most common mistakes is prejudice: our tendency to split the world into "us" and "them" and quickly believe the worst thing possible about whoever "them" is. This is where all our cognitive biases get together and have a bigotry party: we divide the world up according to patterns that might not exist, we make snap judgments based on the first thing to come to mind, we cherry-pick evidence that backs up our beliefs, we desperately try to fit into groups and we confidently believe in our own superiority for no particularly good reason.

(This is reflected in the book in more ways than one: while this is a history of humanity's failures, with a couple of exceptions it's really a history of failures by men; and more often than not, white men. That's because they were often the only ones who were given the chance to fail. Generally it's not a good thing for history books to focus almost exclusively on the deeds of old white guys, but given the subject matter of this one, I think it's probably a fair cop.)

And finally, our desire to fit in with a crowd means that we're extremely prone to fads, crazes and manias of all kinds—brief, flaring obsessions that grip society and send rationality out of the window. These take many different forms. Some can be purely physical, like the inexplicable dancing manias that periodically gripped Europe for about seven centuries in the Middle Ages, in which hundreds of thousands of people would become infected with a sudden and irresistible urge to dance, sometimes to death.

Other manias are financial, as our desire for money combines with our eagerness to be part of a crowd and to believe the stories of whatever get-rich-quick scheme is going around. (In London in 1720, there was such a frenzy of interest in investing in the South Sea that one group of chancers managed to sell stock described as "a company for carrying out an undertaking of great advantage, but nobody to know what it is.") This is how we get financial bubbles—when the perceived value of something far outstrips its actual value. People start investing in the thing not because they necessarily think it has any intrinsic worth, but simply because as long as enough other people think it's worth something, you can still make money. Of course, eventually reality kicks back in, a lot of people lose a lot of money and sometimes the entire economy goes down the pan.

Yet other manias are mass panics, often founded on rumors that play on our fears. That's why witch hunts in one form or another have happened at some point in history in virtually every culture around the world (an estimated 50,000 people died across Europe during the witch manias that lasted from the sixteenth to the eighteenth centuries).

These are just some of the mistakes that recur with wearying predictability throughout the history of human civilization. But, of course, before we could start making them in earnest, we had to invent civilization first.

5 OF THE WEIRDEST MANIAS IN HISTORY

Dancing Manias

Outbreaks of inexplicable, uncontrollable dancing were common in much of Europe between the 1300s and the 1600s, sometimes involving thousands of people. Nobody's entirely sure why.

Well Poisoning

Around the same time, mass panic at false rumors of wells being poisoned were also common—normally blamed on Jews. Some panics led to riots and Jewish homes being burned.

Penis Theft

Outbreaks of panic that malign forces are stealing or shrinking men's penises appear all around the world—blamed on witches in medieval Europe, on poisoned food in Asia or on sorcerers in Africa.

Laughing Epidemics

Since the 1960s, epidemics of unstoppable laughter have occurred in many African schools—one famous outbreak in Tanzania in 1962 lasted a year and a half, forcing schools to temporarily close.

The Red Scare

A classic "moral panic," a wave of anticommunist hysteria swept the USA in the 1940s and 1950s, as the media and populist politicians spread the exaggerated belief that communist agents had infiltrated every part of US society.

2

NICE ENVIRONMENT YOU'VE GOT HERE

Around 13,000 years ago in the Fertile Crescent of ancient Mesopotamia, humans started doing things very differently. They had what you might describe as "a change of lifestyle," and in this case it meant a lot more than cutting down on carbs and joining a gym. Rather than the traditional approach to obtaining food—namely, going to look for it—they hit upon the neat trick of bringing the food to them. They started planting crops.

The rise of agriculture wouldn't just make it easier to grab some lunch; it would completely upturn society and profoundly change the natural world around us. Before agriculture, the standard thing for human groups was to move around with the seasons, following where the food was. Once you've got a load of rice or wheat growing, though, you really need to stick around to look after it. And so you get permanent settlements, villages and, sometime after that, towns. And, of course, all the stuff that goes with that.

Agriculture was such an obviously great idea that it sprang up independently in loads of different places, all within a few thousand years of each other on several different continents—in Mesopotamia, India, China, Central America and South America at the very least. Except that there's a school of thought that says agriculture wasn't actually our greatest leap forward. In fact, it may have been a dreadful, dreadful mistake.

For starters, the origin of agriculture was also the origin of the fun concept of "wealth inequality," as elites began to emerge who had way more stuff than everybody else and started bossing everybody else around. It may also have been the origin of war as we know it, because once you have a village, you also have the danger of raids on it by the next village. Agriculture brings new diseases into contact with humans, while living together in ever larger settlements creates the conditions for epidemics. There's also evidence that suggests people in nonagricultural societies ate more, worked less and may well have been healthier.

Basically (this idea goes), an awful lot of what sucks about modern life was because thousands of years ago somebody stuck some seeds in the ground. Agriculture hung around not because it made everybody's lives better, but because it gave societies that did it a Darwinian leg up over the ones that didn't: they could have more children faster (agriculture can feed more people, and once you're no longer moving around all the time, you don't have to wait for your kid to be able to walk before having another), and they could claim more and more land, eventually chasing all the nonfarmers off. As the author Jared Diamond, a proponent of the "agriculture was a horrible mistake" theory, put it in a 1987 article in *Discover* magazine: "Forced to choose between limiting population or trying to increase food production, we chose the latter and ended up with starvation, warfare and tyranny." In short, we went for quantity over quality. Classic humans.

But in addition to all…this [vaguely waves hand around at the state of the world], agriculture started us on a path that would lead to many more direct, more dramatic screw-ups. The dawn of agriculture was when we started to change the environment around us—after all, that's what farming is. You take plants and put them in places they weren't planning on being. You start to reshape the landscape. You try to get rid of the things that you don't want so you can fit in more of what you do want.

Anyway, it turns out that we're *really* bad at thinking that kind of stuff through.

The world around us right now is profoundly different from the one our ancestors first planted seeds in 13,000 years ago. Agriculture has altered the landscape and transplanted species across continents, while cities and industry and our natural tendency to just throw away any garbage we don't want has changed the soil, the sea and the air. And without wanting to get all We Must Not Anger the Earth Mother on you, sometimes the natural world is not here for our bullshit.

That's what famously happened on the central plains of the USA in the first half of the twentieth century. As is often the case, to begin with everything was going pretty swell. America was expanding westward, and people were living out a version of the American Dream. Government policies encouraged people to move west and get farming, with settlers being granted free plots of land across the Great Plains. Unfortunately, by the beginning of the century most of the good farming land—basically the bits with a decent water supply—had already been claimed. People were understandably slightly less enthusiastic about heading out to farm dry, dusty land, so the government doubled the amount of dry, dusty land they'd be given. "Sounds like a good deal," the settlers said.

If that drive to farm every scrap of land doesn't seem, with hindsight, like the greatest idea in the world, there were a bunch of reasons why people assumed it would be fine. There was the romantic—the nostalgic appeal of an agrarian nation of pioneers—and the pragmatic, a growing country's basic need for food. But there was also some highly dodgy science, bordering on religion: the theory that "rain follows the plow," that the simple act of starting to farm land would summon rain clouds, turning the desert fertile and verdant. Under this theory, the only thing stopping the expansion of farming in America was a lack of will. It's like that Kevin Costner movie, but with cereal crops instead of ghosts playing baseball. If you farm it, rains will come.

They really believed in this, so it almost seems mean to point out that the reason it so often started raining as soon as farmers moved into an area was that the middle of the nineteenth century, when the theory was developed, was simply an unusually rainy period. Those rains, unfortunately, would not hang around forever.

Come World War I and suddenly all that farmland seemed like a great idea: Europe's food production had ground to a halt, but America was able to pick up the slack. Prices were sky-high, the rains were good and the government kicked in some generous subsidies to farmers for planting wheat, so naturally farmers did, plowing up ever more prairie as they did so.

After the war, though, wheat prices fell dramatically. Now, if you're a wheat farmer, and you're not making enough money from your wheat, then the solution is obvious: you need to plant more wheat. Farmers invested in new mechanical plows, tearing up even more of the soil. Even more wheat meant even lower prices, which...and so on.

Then, suddenly, the rains didn't come. The soil dried out, and the roots of the grasses that had held the topsoil together through previous droughts suddenly weren't there. The soil turned to dust, and the wind picked that dust up into enormous, roiling clouds.

It's those fearsome dust storms—the "black blizzards" that blotted out the sun, choked the air and reduced visibility to a few feet—that became the symbol of the Dust Bowl. In the worst years, hardly a day would go by in summer without the storms whipping up, and even when the wind died, the dust clouds would still hang in the sky. Sometimes residents barely saw the sun for days. The dust storms had astonishing range, with some of them traveling thousands of miles, blanketing cities like Washington, DC, and New York in a thick soil smog, and coating ships hundreds of miles off the East Coast with a fine layer of dust.

The drought and the dust storms persisted for almost a decade. It was economically ruinous, and several million people were forced to abandon their homesteads. Many never returned, settling instead even farther west, a large number in California. Some of the land never fully recovered, even when the rains came back.

The American Dust Bowl is one of the more famous examples of the unintended consequences of messing around with our environment. But from mass-scale geo-engineering to tiny plastic beads, from deforestation to rivers doing things that rivers definitely shouldn't do, it's not the only one.

Take the Aral Sea, for example—although you'll have to move quickly, because there's not a lot of it left to take.

The Aral Sea, despite its name prominently involving the word *sea*, is not actually a sea. Instead, it's a saltwater lake, though a very, very large one—at over 26,000 square miles, one of the largest lakes in the world, or at least it was. The problem, you see, is that it is not 26,000 square miles anymore.

A dust cloud in Colorado during the American Dust Bowl, 1936

It's now about 2,600 square miles, although that kind of goes up and down. Once almost the size of Ireland, it's down to a tenth of its former size, and it's lost over 80 percent of its water. It's also not a single huge lake anymore—it's now, roughly, four much smaller ones. "Roughly," because one of the lakes may have vanished entirely. What little is left of the Aral Sea is now virtually dead, a lifeless ghost sea surrounded by the rusting and decaying skeletons of long-stranded ships that are now miles from any water.

Which prompts the question: How, exactly, do you lose an entire bloody sea? (Well, an entire big lake.)

The simple answer is: you divert the two rivers that used to flow into it, because you've had the bright idea of growing cotton in the desert. That's what the Soviet authorities did from the 1960s onward, because they really wanted to have more cotton. So they undertook a huge project to redirect water from the Amu Darya (which flowed into the Aral Sea from Uzbekistan) and the Syr Darya (which reached the sea through Kazakhstan) so that the bone-dry plains of the Kyzylkum Desert could be converted into a monoculture that would supply the Soviet

Union's cotton needs. Now, in fairness, the plan to irrigate the wannabe farms of Turkmenistan, Kazakhstan and Uzbekistan was a partial success—albeit a hugely wasteful one, because the thing about deserts is that they're quite dry and absorbent, so as much as 75 percent of the diverted river water never even made it to the farms. (There was also the issue of the defoliant chemicals used on the cotton, which caused sky-high rates of infant mortality and birth defects.)

But while this was good-ish news for the infant cotton industry of Central Asia, it was devastating for the Aral Sea and its surroundings. It doesn't seem to have occurred to anybody—or they simply didn't care—that when you stop water flowing into a lake, you quickly end up with much less lake.

Immediately from the sixties, but at an accelerating pace from the late eighties until the present day, the Aral Sea began to shrink. Only about a fifth of the water that had flowed into it before came from rainfall, with the rivers supplying the rest. So once they were virtually gone, there wasn't enough water left to replace that lost to evaporation. The water level started dropping, and new islands and isthmuses began appearing; by the turn of the millennium, the lake had split into two: a small northern section and a larger southern section with a massive island in the middle of it. The waters kept falling, so that the island kept growing until only a tiny ribbon of water connected the eastern and western halves of the southern sea. Eventually they split, too, and then in the summer of 2014, satellite photographs revealed that the eastern section had dried up entirely, leaving only desert in its place. That eastern lake now just kind of comes and goes, depending on what the weather's like.

This would have been bad enough, but the thing when a lake disappears is that all the stuff in the water…doesn't. Salt, particularly. While the Aral Sea receded, the salt continued to hang around, making the waters saltier and saltier and less and less capable of supporting life. The salt density rocketed by a factor of ten, which killed virtually every living thing in the lakes, destroying a thriving fishing industry that had supported 60,000 jobs. Not just that, but the pollutants from industry and agriculture became more concentrated, and then got laid down on the exposed surface of new lands emerging as the waters drew back. Deserts being what they are, the wind then picked up tons and tons of toxic dust and salt from the newly arid landscape and dumped it right on the villages and towns surrounding the former lakes where millions of people lived. Respiratory disease and cancer shot up.

It's not necessarily all over for the Aral Sea; recent (very expensive) efforts to divert some water back to it have led to a bit of an improvement in the small northern sea, with fish stocks gradually returning, although the southern sea is pretty much a write-off. But it still stands as a testament to our capacity for thinking we can just make large-scale changes to the geography of our environment without there being some sort of blowback.

Weirdly, this isn't even the first time this has happened to one of the rivers involved. I'm not sure if there's a world record for "most consistently diverted river," but the Amu Darya has to be in with a shout. For centuries, interventions by both nature and a series of human regimes have repeatedly changed its course from flowing to the Aral Sea to the Caspian Sea (or sometimes both) and back again. In the second century CE it's thought to have flowed into the desert, where it evaporated, before at some point switching to the Aral Sea. In the early thirteenth century, a particularly drastic intervention by the Mongol Empire changed its course again (more on that in a later chapter), sending at least part of it toward the Caspian, before it returned to the Aral Sea sometime before the 1600s. In the 1870s, long before the Soviet Union came into being, the Russian Empire gave serious consideration to diverting it back to the Caspian, on the grounds that they thought the fresh water was going to waste pouring into a salty lake. Which…is not how it works, guys.

It was agriculture that first led us to change the environment in dramatic ways, often with unforeseen consequences, but it's not the only way we do that anymore. Agriculture has in many ways been outpaced by the rise of industry, and human beings' apparently unquenchable desire to dump stuff we don't want into the environment without really thinking about the consequences.

An example of those consequences would be the way that, late one morning on a warm summer's day in 1969, the Cuyahoga River caught fire.

Rivers, to be clear, aren't meant to do that. For any readers who are still vague on the general concept of rivers, they're medium to large natural channels of flowing water, and water is not commonly regarded as being especially flammable. Rivers do an awful lot of things—transporting water from high ground to lowlands, providing a metaphor for the passage of time, forming oxbow lakes so children can at least remember something from their geography lessons—but bursting into flames is absolutely not supposed to be one of them.

The Cuyahoga River did, though, and what's more it wasn't the first time it had done it. Not even close. In fact, the Cuyahoga—which meanders slowly through industrial northern Ohio before bisecting the city of Cleveland and plopping into Lake Erie, and which was described by one nineteenth-century mayor of Cleveland as "an open sewer through the center of the city"—was so polluted that it had caught fire no fewer than 13 times in the previous 101 years. It burned in 1868, 1883, 1887, 1912 (when five men died in the resulting explosions), 1922 and 1930. In 1936, the fire was so bad that it raged for five days—which, to reiterate, is not traditional behavior for rivers. It burned again in 1941 and in 1948 and then most destructively in 1952, when the two-inch-thick slick of oil that lay across its surface lit up, sparking an enormous conflagration that destroyed a bridge and a shipyard and caused as much as $1.5 million worth of damage.

In comparison to 1952, the 1969 blaze was comparatively small stuff. Caused by the ignition of a coagulated mix of oil, industrial waste and debris that had combined to form a sort of flammable trashberg floating down the river, it put on an impressive show (the flames were five stories high) but was under control inside half an hour, the Cleveland fire department clearly being on top of their river-fire-extinguishing game by this point. The people of the city were apparently so used to this sort of stuff that *the goddamn river catching fire* only merited five paragraphs of story deep on the inside pages of the *Cleveland Plain Dealer.*

But if Cleveland's long-suffering populace were kind of "oh, that again" about river fires by 1969, the nation as a whole was not. Things had changed since the last time the Cuyahoga burned. It was the sixties, after all, and society was being shaken to its core by a series of revolutionary new ideas, such as "having fewer wars," "not being quite as racist" and "maybe trying not to fuck the planet up entirely."

So when *Time* magazine cottoned on to the fire a few weeks later, they wrote it up in a story on the state of the nation's rivers titled "America's Sewage System and the Price of Optimism," which included this memorable description of the Cuyahoga: "Chocolate-brown, oily, bubbling with subsurface gases, it oozes rather than flows...an open sewer filling Lake Erie with scummy wavelets." *Time*'s article grabbed the country's attention and prompted widespread demands for change—largely thanks to the jaw-dropping picture that accompanied the story, a dramatic shot of a boat engulfed in the flames of the burning river as fire crews struggled to contain the blaze. Actually, that picture wasn't of the 1969 fire; it was an archive photograph from the 1952 blaze, because the 1969 fire had been dealt with so quickly that it was out before any photographers or film crews managed to turn up. The picture hadn't caught the national imagination back in 1952, but now it worked a treat. Sometimes timing is everything.

1952 Time *magazine photo of a boat engulfed in flames on the Cuyahoga River, Cleveland, Ohio*

Since the 1800s, the industries of Ohio had been cheerfully spilling both the by-products and indeed the products of their work into the Cuyahoga. This led to regular bouts of the media, politicians and public saying things like, "Uh, maybe we should do something about this?" followed by nobody really doing anything about it. A few half-hearted measures were implemented in the years after the war, but they were mostly concerned with making the river safe for shipping rather than making the river not inherently flammable.

Still, it was perhaps slightly unfair that the Cuyahoga became the national symbol of humanity's inaction in the face of environmental destruction, if only because the year before the fire, the city of Cleveland had actually passed some laws to finally get the river cleaned up. Quite a few local officials sounded a bit miffed at the fact that they had become the poster river for the filthy state of the nation's waterways (to the extent that there were even songs written about it). "We were already doing the things we needed to clean up things there, and then the fire happened," one said plaintively.

After all, theirs wasn't even the only burning river in the country at the time. The Buffalo River caught fire in 1968, the year before Cuyahoga, while the Rouge River in Michigan burst into flames just a few months after it in October 1969. ("When you have a river that burns, for crying out loud, you have troubles," the *Detroit Free Press* lamented in the aftermath.) Cuyahoga wasn't even the only river in the US to have caught fire on multiple occasions—in the nineteenth century, the Chicago River saw conflagrations with enough regularity that the community would come out and watch like it was a Fourth of July display—although it definitely takes the prize for Most Consistently on Fire River, North American Category.

Still, the tale of the flaming river did its job, spurring national action. The nascent environmental movement, already primed by books like Rachel Carson's 1962 work *Silent Spring*, began to coalesce (the first Earth Day was marked the following year). Congress was forced to take action, passing the Clean Water Act in 1972. Gradually the state of America's waterways improved, to the point where they now hardly ever catch fire. In a rare example of a happy ending in this book, people actually did what they needed to do to make things better, and hahaha, there's absolutely no chance that the Trump administration would ever attempt to overturn clean water standards because they're worried industries aren't allowed to pollute rivers enough. [Puts finger to ear.] Oh, I'm being told that's exactly what they've done.

Large bodies of water erupting into flames may be one of the more dramatic examples of humanity's unerring ability to take the natural world around us and make it worse, but they're hardly alone. The world is full of examples of how we've managed to make a huge mess basically everywhere we've gone. Did you know there's a vast "dead zone" in the Gulf of Mexico? It's a massive plume of mostly ruined sea, spreading out from where the fertilizer running off the agricultural lands of the southern

USA has caused great algal blooms, the algae going hog-wild and robbing the water of oxygen, killing everything else that isn't algae. Good work, guys!

Or how about our fondness for just throwing stuff away without really considering that it has to end up somewhere, which has given rise to the massive electronic wasteland of Guiyu in China, an infamous 20-square-mile graveyard of the world's unwanted gadgets, piled high with outdated laptops and last year's smartphones. Technically Guiyu is in the business of recycling, which is good! Unfortunately (until recently), it was also hell on earth, with thick plumes of black smoke filling the air, toxic heavy metals leaching into both the soil and the people after the scrap is washed with hydrochloric acid, and the smell of burning plastic everywhere. (That was until the Chinese government cracked down on it in the past few years, enforcing higher health and safety standards—after which one resident told the *South China Morning Post* that the air quality was much improved. "You can only smell the burning of metal when you are really close," Yang Linxuan said.)

An electronic waste mountain in Guiyu, China

Perhaps our most impressive piece of work is the Great Pacific Trash Vortex. It's almost poetic that in the middle of the ocean there's a vast, swirling rubbish dump of crap we've casually disposed of—an area the size of Texas where the ocean currents of the North Pacific Gyre keep our waste products endlessly circling around the sea. Mostly made from microscopic particles of plastic and fragments of discarded fishing equipment, it's invisible to the naked eye, but for marine life it is a very real thing. Scientists recently estimated that since we started the widespread use of plastic in the 1950s, we've made over 8,300 million tons of it. Of that, we've thrown away 6,300 million tons, which is now just hanging around the surface of the earth. Yay, humans.

But if you want the most poignant example of how humans can destroy their own habitat without really meaning to, you need to look to an island covered in massive stone heads.

Heads You Lose

When the first Europeans landed on Easter Island in 1722 (a Dutch expedition that was looking for a supposedly undiscovered continent that doesn't exist, the idiots), they were baffled. How could this tiny, profoundly isolated Polynesian nation, lacking modern technology or any trees, possibly have erected the vast, elaborate statues—some of them 70 feet tall and weighing almost 90 tons—that covered major parts of the island?

Obviously, the Netherlanders' state of curiosity didn't last long: they quickly set about doing their regular European thing, namely shooting a bunch of the local inhabitants dead after a series of misunderstandings. Over the next few decades, further European visitors did more of the stuff that Europeans tended to do in places they'd just "discovered," like introducing deadly new diseases, kidnapping the local population into slavery and generally patronizing the hell out of them. (See the later chapter on colonialism.)

Over the following centuries, white people would come up

with a bunch of theories about how those mysterious statues could possibly have appeared on an island full of "primitive" people—mostly involving implausible ocean crossings from faraway continents, or sometimes aliens. ("Aliens must have done it" is a remarkably popular and obviously extremely rational solution to the conundrum of nonwhite people building things that white people can't imagine them having built.) The answer to the question is, of course, obvious: the Polynesians put them there.

At the time they first landed on Rapa Nui (to give it its local name), the Polynesians were one of the great world civilizations who had explored and settled on islands across thousands of miles of ocean. Meanwhile, a few stray Vikings aside, Europeans hadn't really got out of their backyards.

Rapa Nui was home to an advanced culture, with intergroup cooperation, intensive agriculture, a socially stratified society and people commuting to work: basically all of the bullshit that we generally associate with being fancy and proper. The statues—*moai* in Polynesian—were the crowning achievement of an art form common to other Polynesian societies. They were important to Rapa Nui's society for both spiritual and political reasons, honoring the ancestors whose faces they depicted, while functioning as symbols of prestige for those who ordered their construction.

So the puzzle turned into a different one: not how did the statues get there, but instead, where had all the trees gone? Because however the Rapa Nui people got those statues into place, they'd have needed a ton of big-ass logs to do it. And how did the mighty civilization that put them there turn into the small society of subsistence farmers with threadbare canoes that greeted (and then got shot by) those first Dutch sailors?

The answer is that the Rapa Nui people both got unlucky, and fucked up.

They got unlucky because, it turns out, their island's geography and ecology was unusually vulnerable to the effects of deforestation. As Jared Diamond (he of the "agriculture was our biggest mistake" theory) explains in his book *Collapse*—which puts the people of Rapa Nui front and center—compared to most Polynesian islands, Easter Island is small, dry, flat, cold and remote: all things that make it less likely that the trees you cut down will get naturally replaced.

And they fucked up because, in their efforts to keep building better houses and better canoes and better infrastructure to move statues into place, they kept cutting the forest down, maybe not realizing that those trees wouldn't come back, until suddenly there were no more left. It was the tragedy of the commons writ large. Nobody cutting down any single tree was responsible for the problem, up until it was too late: at which point everybody was responsible.

The effects were devastating to Rapa Nui's society. Without the trees, they couldn't make the canoes that let them fish in the open ocean; the rootless and unprotected soil started to erode in the wind and rain, becoming infertile and causing landslides that wiped out villages; in the cold winters, they were forced to burn much of the remaining vegetation to stay warm.

And as things got worse, competition between groups for increasingly scarce resources increased. This seems to have led to an outcome that was tragic but weirdly predictable, given how people often act in a desperate situation where they're hungry for social standing, or morale, or just a bit of reassurance that they haven't made a terrible mistake. They didn't stop. In fact, they doubled down. The people of Rapa Nui appear to have thrown themselves into building ever larger and larger statues, because...well, that's pretty much what humans always do when faced with a problem that they're worried they can't solve. The last statues carved on the island never even made it out of the quarry, while others lie toppled by the wayside, never having reached their destination as the whole project collapsed.

The Polynesians weren't any less clever than you or me; they weren't primitive or unaware of their environment. If you think that a society faced with potential environmental disaster ignoring the problem and doing even more of the stuff that caused it in the first place sounds foolish, then, er, hi. Maaaaaaybe take a little look around you? (Then please turn your thermostat down and take out the recycling.)

In *Collapse*, Jared Diamond ponders the question: "What did the Easter Islander who cut down the last palm tree say while he was doing it?" Which is a very good question, and one that's pretty hard to answer. Possibly it was some Polynesian version of YOLO.

But maybe a better question might be what the Easter Islander who cut down the second-to-last tree, or the third-to-last, or the fourth-to-last was thinking. If the rest of human history is any guide, there's probably a fairly decent chance he was thinking something along the lines of "Not my problem, mate."

7 AMAZING SIGHTS YOU'LL NEVER SEE, BECAUSE HUMANS RUINED THEM

The Parthenon

One of the jewels of Ancient Greece, until in 1687 the Ottomans used it as a gunpowder store during a war with Venice. One lucky Venetian shot later—no more Parthenon.

Temple of Artemis

One of the actual Seven Wonders of the Ancient World, until 356 BCE, when a bloke called Herostratus burned it down because he wanted attention.

Boeung Kak Lake

The largest and most beautiful lake in the Cambodian capital of Phnom Penh, until it was decided to pump it full of sand to build luxury apartments on it. Now a puddle.

Buddhas of Bamiyan

The magnificent statues of Gautam Buddha in central Afghanistan, over a hundred feet tall, were blown up by the Taliban in 2001 because they were "idols." FFS.

Nohmul

A great Mayan pyramid, the finest Mayan remains in Belize, was torn down in 2013 by some building contractors because they wanted gravel for nearby roadworks.

Slims River

A vast river in Canada's Yukon territory that completely vanished in the space of four days in 2017, as climate change caused the glacier it flowed from to retreat.

Ténéré Tree

Famously the most isolated tree on the planet, alone in the middle of the Sahara Desert—until 1973, when despite it being the only tree for 250 miles, a drunk driver still managed to drive his truck into it.

3

LIFE, UH,
FINDS A WAY

Alongside the development of planting crops, the first farmers all those thousands of years ago started to do something else that would change our world in strange and unpredictable ways—they began domesticating animals.

Actually, the very first domesticated animal almost certainly predates the development of agriculture by thousands of years—although that may have been more of a happy accident than a clever plan. Dogs were the original domestic animals, and they seem to have become domesticated sometime between 40,000 and 15,000 years ago, in either Europe, Siberia, India, China or somewhere else (the uncertainty comes from the fact that dog

DNA is a bit of a mess, because dogs will cheerfully shag pretty much any other dog they come across). While it's possible that this happened because an enterprising hunter-gatherer ancestor of ours woke up one day and said, "I am going to make friends with a wolf and he will be a Very Good Boy," it's more likely that dogs were (at the start, at least) basically self-domesticating. The most plausible dog origin story is simply that wolves started following humans around, because humans had food and tended to discard their leftovers. Over time, those wolves started to adapt more and more to life with humans; meanwhile, humans began to realize that having some friendly wolves living with them was actually pretty useful for protection and hunting, and also they were all fluffy-wuffy, yes they were.

But once agriculture kicked off in earnest, humans began to work out that the thing they were doing with plants might also work with animals, and would save everybody the bother of going out hunting. Around 11,000 years ago, goats and sheep were domesticated in Mesopotamia. Five hundred years after that, cattle were domesticated in modern-day Turkey, and then again in what is now Pakistan. Pigs were also domesticated twice, about 9,000 years ago—in China and again in Turkey. In the Eurasian Steppe, probably somewhere around Kazakhstan, horses were domesticated between 6,000 and 5,500 years ago. Meanwhile, in Peru, around 7,000 years ago, humans first tamed the guinea pig. Which, granted, sounds slightly less impressive but honestly was pretty cool.

Domestication of animals had a lot of useful upsides—a ready supply of protein, wool for clothing and manure for fertilizing crops. Of course, it wasn't all good news, as we mentioned in the previous chapter. Keeping animals at close quarters makes it a lot easier for diseases to jump from animals to humans; keeping horses and cows seems to have been linked with the origin of wealth inequality; and the military uses of horses and elephants made war a lot more...warlike.

In addition, the domestication of animals gave us a very clear idea that we were the masters of nature, and that from now on animals and plants would do our bidding. Unfortunately, as we'll see in this chapter, that's not exactly how it always works out. Humans' persistent belief that we can make living things do exactly what we want them to has a rather nasty habit of backfiring on us.

For example, let's rewind to the year 1859, when Thomas Austin was feeling a bit homesick.

Austin was an Englishman, but had arrived in the colony of Australia as a teenager. Now, a couple of decades later, he was a prosperous landowner and sheep farmer, presiding over a vast spread of 29,000 acres at his home near Victoria. There he replicated the pursuits of his ancestral lands with enthusiasm: a keen sportsman, he bred and trained racehorses and turned much of his acreage into a preserve for wildlife and hunting. His estate gained such a reputation in Australian high society that the Duke of Edinburgh was a regular visitor on his trips to Australia. When Austin died decades later, his glowing obituary said that "a better representative of the real old English country gentleman could not be found, either here or at home."

His determination to live the life of a traditional country squire on the far side of the world led him to do everything in his power to replicate a little bit of England in the Antipodes. And that, unfortunately, is where it went to shit.

That's because Austin decided that his hunting would be vastly improved with the importation of some classic English animals to shoot (wallabies, presumably, just didn't quite cut it for him). He had his nephew ship over pheasants and partridges, hares and blackbirds and thrushes. And crucially, he imported 24 English rabbits. "The introduction of a few rabbits," he said, "could do little harm and might provide a touch of home in addition to a spot of hunting."

He was very, very wrong about the "little harm" bit. Although in fairness he was right that they would indeed provide a spot of hunting.

Austin wasn't the first person to bring rabbits to Australia, but it was his rabbits that were largely responsible for the catastrophe that was about to strike. The thing about rabbits is that they breed like...well, rabbits. The scale of the problem should probably have been evident from the fact that in 1861, just a couple of years after Austin's initial shipment arrived, he boasted in a letter that "English wild rabbit I have in thousands."

It didn't stay in the thousands. A decade after Austin introduced them, two million rabbits were being shot each year in Victoria without denting their population growth in the slightest. The rabbit army soon spread all across Victoria, moving at an estimated 80 miles a year. They were seen in New South Wales by 1880, in South Australia and Queensland by 1886, Western Australia by 1890 and the Northern Territory by 1900.

By the 1920s, at the height of the rabbit plague, Australia's rabbit population was estimated at 10 billion. There were 3,000 of them for every square mile. Australia was quite literally covered in rabbits.

The rabbits didn't just breed; they ate (breeding is hungry work, after all). They stripped the land bare of vegetation, driving many plant species into extinction. The competition for food brought a number of Australian animals to the brink of extinction, as well, while without plant roots to hold the soil together the land itself crumbled and eroded.

The scale of the problem was clear by the 1880s, and authorities were at their wits' end. Nothing they tried seemed to be capable of stopping the floppy-eared onslaught. The government of New South Wales placed a slightly desperate-sounding advertisement in the *Sydney Morning Herald*, promising to pay "the sum of £25,000 to any person or persons who will make known...any method or process not previously known in the Colony for the effectual extermination of rabbits."

A horde of rabbits drinking at a watering hole in Adelaide, Australia, 1961

Over the following decades, Australia tried shooting, trapping and poisoning the rabbits. They tried burning or fumigating their warrens or sending ferrets into the tunnels to flush them out. In the 1900s, they built a fence over a thousand miles long to try and keep the rabbits out of Western Australia, but that didn't work because it turns out that rabbits can dig tunnels and, apparently, learn to climb fences.

Australia's rabbit problem is one of the most famous examples of something that we've only figured out quite late in the day: ecosystems are ridiculously complex things and you mess with them at your peril. Animals and plants will not simply play by your rules when you casually decide to move them from one place to another. "Life," as a great philosopher once said, "breaks free; it expands to new territories and crashes through barriers—painfully, maybe even dangerously. But, uh, well, there it is." (Okay, it was Jeff Goldblum in *Jurassic Park* who said that. As I say, a great philosopher.)

Ironically, after the initial fuck-up of introducing rabbits into

Australia in the first place, the eventual solution was also a fuck-up. For several decades Australian scientists had been experimenting with using biological warfare on the rabbits: introducing diseases in the hope that they'd be killed off, most famously myxomatosis in the 1950s. That worked pretty well for a while, reducing the rabbit population dramatically, but it didn't stick. It relied on mosquitoes to transmit the virus, so wasn't effective in areas where mosquitoes wouldn't breed, and eventually the surviving rabbits developed resistance to the disease and numbers started climbing again.

But the scientists carried on researching new biological agents. In the 1990s, they were working on rabbit hemorrhagic disease virus. Now, experimenting with diseases is a dangerous business, and so the scientists were doing their work on an island off the south coast, to reduce the risk of the virus getting loose and spreading to the mainland. Go on. Guess what happened.

Yep, in 1995, the virus got loose and spread to the mainland. Life broke free, in this case by hitching a ride on some flies. But having accidentally released a deadly (to rabbits) pathogen into the wild, the scientists were rather pleased to note that…it seemed to be working. In the twenty years since rabbit hemorrhagic disease virus was mistakenly released into the wild, rabbit populations in South Australia have declined again, while vegetation has returned and the many animals that had been pushed to the brink of extinction have seen their numbers surge back. Let's just hope that rabbit hemorrhagic disease virus doesn't turn out to have any other side effects.

Australia's rabbits are far from alone in proving that sometimes we should leave animals and plants where we found them.

Like the Nile perch, a six-foot-long ravenous predator that, as you might guess from the name, comes from the Nile. However, the British colonizers of East Africa had bigger plans for it. They thought it would be a terribly good idea to introduce it into Lake Victoria, the largest lake in Africa. Lake Victoria

A man carries an 80-kilogram Nile perch in Uganda

already had lots of fish in it, and local fishermen were perfectly content fishing those fish, but the British thought that this situation could be improved. The biggest group of fish in the lake at the time were hundreds of different species of cichlids, those small, adorable-looking fish beloved of aquarium keepers. Unfortunately for the cichlids, the British colonial officials hated them, describing them as "trash fish."

They decided that Lake Victoria would be much better with bigger, cooler fish in it. It would make for superior fishing, they reckoned. Lots of biologists warned them that this was not a great idea, but in 1954 they went ahead and introduced the Nile perch into the lake. The Nile perch then did what Nile perch do: they ate their way through species after species.

The British officials were right about one thing, in that it really did make for superior fishing. The fishing industry boomed, with Nile perch proving immensely popular both as a commercial catch for food and an enjoyable catch for sport. But while the value of the fishing industry shot up by 500 percent, supporting hundreds of thousands of jobs, the number of species in Lake Victoria plummeted. More than 500 other species became extinct, including over 200 species of the poor unfortunate cichlids.

It's not just animals that can get out of control. Kudzu, a vine common across Asia, was widely introduced into the USA in the 1930s in an attempt to solve a problem that we've already mentioned: the Dust Bowl. Officials hoped that the fast-growing vine would help knot the soil back together and prevent further erosion. And it was quite good at that. Unfortunately, it was also quite good at smothering other plants and trees, as well as houses, cars and anything else it came across. It became so widespread across the southern United States that it was given the nickname "the vine that ate the South."

In fairness to kudzu, it isn't quite the triffid-like demon plant that some mythology suggests, and recent studies have found it covers less land than commonly thought. Still, there's an awful lot of it where 80 years ago there wasn't any of it, and it remains officially listed by the US government as a "noxious weed."

But now might be the time to start feeling sorry for it, because the invasive species has gained an invasive species of its own. Sometime in 2009, the Japanese kudzu bug managed to make its way across the Pacific, and must have been delighted on landing in Atlanta to discover that there was a load of kudzu already there for it to eat. In the space of three years, it had spread through three states, wiping out as much as a third of the kudzu's biomass. In case you're thinking, well, that's good, kudzu problem solved, it's unfortunately not quite that simple: kudzu bugs also destroy soy crops, a major source of income in many of the affected states. The accidental solution to one problem might turn out to be a much bigger problem in its own right.

Our apparent desire to introduce new species where they've no right to be doesn't even stop at species that already exist: sometimes we manage to create whole new species. That's what happened in 1956 when the Brazilian scientist Warwick Estevam Kerr imported some African queen bees from Tanzania, in an effort to cross-breed them with European bees—the hope being that their combined traits would produce a species better suited to the Brazilian environment.

Unfortunately, after a year of breeding experiments, the thing that was always going to happen happened. A beekeeper working at Kerr's lab in Rio Claro, a city to the south of São Paulo, had a very bad day on the job. Twenty-six of the Tanzanian queen bees escaped, followed closely by their personal swarms of European bees, and set up home in Brazil. The queens started breeding indiscriminately with any male bees they came across, producing hybrid strains with several different species. These new "Africanized" bees started spreading rapidly across South America, then Central America and then into the USA. They're actually smaller and have less venom than the bees that came before them, but they are way more aggressive at defending their hives—producing up to ten times the number of stings. As many as a thousand people have died as a result of those stings, which is why the bees have ended up with the nickname "killer bees." Which is a bit unfair. They're just misunderstood.

But in the annals of humans learning the very hard way that ecosystems are complicated and that messing with the delicate balance of nature will come back to bite you, two stories stand out above all else. On different sides of the world, several decades apart, a fanatical dictator and an eccentric literature lover made mirror-image mistakes that had profound consequences. Both of their mistakes came from the same source: they radically underestimated birds.

Don't Underestimate Birds, Part I: A Pest Too Far

Mao Zedong's Four Pests campaign has to rank as the most disastrous entirely successful public health policy ever. It pulled together every part of society to meet its goals, which it surpassed to an astonishing degree—and half of those goals almost certainly resulted in major widespread improvements to the health of the nation. Two out of four isn't bad, you might think.

The trouble is that the fourth goal resulted in tens of millions of deaths.

The problem stems from that same failure to realize that ecosystems are complicated and unpredictable. Oh yeah, let's just add a species here, maybe trim a couple of species there, we think. That'll make everything better. At which point Unintended Consequences rocks up with her pals Knock-On Effects and Cascading Failure and throws a hubris party.

When Chairman Mao's communists took power in China in late 1949, the country was in the grip of a medical crisis. Infectious diseases, from cholera to plague to malaria, were running rife. If Mao's goal of rapidly transforming the country from a largely agrarian nation only a few decades out of feudalism into a modern industrial powerhouse were to be met, something would have to be done.

Some of the solutions were obvious and sensible—mass vaccination programs, improved sanitation, that sort of thing. The problems started when Mao decided to focus on blaming animals for the country's woes.

Mosquitoes spread malaria, rats spread plague; that much was pretty undeniable. And so a nationwide plan to reduce their numbers was hatched. Unfortunately, Mao didn't stop there. If it had just been a Two Pests campaign, then things might have worked out okay. But Mao decided (without bothering to do anything like, you know, ask experts their opinion or anything) to add in two other species, as well. Flies were to be wiped out, on the grounds that flies are annoying. And the fourth pest? Sparrows.

The problem with sparrows, the thinking went, was that they ate grain. A single sparrow could eat as much as 4.5 kilograms of grain every single year—grain that could be used instead to feed the people of China. They did the math and determined that 60,000 extra people could be fed for every million sparrows that were eliminated. Who could argue with that?

The Four Pests campaign began in 1958, and it was a remarkable effort. A countrywide poster campaign demanded that every citizen, from the youngest to the oldest, do their duty and kill the shit out of as many animals as possible. "Birds," it was declared, "are public animals of capitalism." The people were armed with everything from flyswatters to rifles, with schoolchildren being trained in how to shoot down as many sparrows as possible. Jubilant sparrow-hating crowds took to the streets waving flags as they joined battle with the birds. Sparrows' nests were destroyed and their eggs smashed, while citizens banging pots and pans would drive them from trees so they could never rest until, exhausted, they fell dead from the sky. In Shanghai alone, it was estimated that almost 200,000 sparrows died on the first day of hostilities. "No warrior shall be withdrawn," the *People's Daily* wrote, "until the battle is won."

The battle was, indeed, won. In terms of achieving its stated goals, it was a triumph—an overwhelming victory for humanity against the forces of small animals. In total, the Four Pests campaign is estimated to have killed 1.5 billion rats, 11 million kilograms of mosquitoes, 100 million kilograms of flies...and a billion sparrows.

Unfortunately, it quickly became apparent what the problem with this was: those billion sparrows hadn't just been eating grain. They'd also been eating insects. In particular, they ate locusts.

Suddenly freed from the constraints of a billion predators keeping their numbers down, the locusts of China celebrated like it was New Year every day. Unlike sparrows—who'd eat a bit of grain here and there—the locusts tore through the crops of China in vast, relentless devouring clouds. In 1959, an actual expert (ornithologist Tso-hsin Cheng, who had been trying to warn people how bad an idea this all was) was finally listened to, and sparrows were replaced on the list of official pests-we-want-to-kill by bedbugs. But by then it was too late; you can't just replace a billion sparrows on a whim once you've wiped them out.

To be clear, the destruction of the sparrows wasn't the only cause of the great famine that struck China in the years between 1959 and 1962—a perfect storm of terrible decisions helped to cause it. A Party-mandated shift from traditional subsistence farming to high-value cash crops, a suite of destructive new agricultural techniques based on the pseudoscience of the Soviet biologist Trofim Lysenko and the central government appropriating all produce and diverting it away from local communities each played its part. Incentives that pushed officials at every level to report positive results led to the delusion on the part of the country's leaders that, basically, Everything Was Fine and the nation had more than enough food. This meant that when several years of terrible weather hit (flooding in some parts of the country, drought in others), there were no reserves to see them through.

But all that sparrow-murdering, and the subsequent obliteration of crops by the real pests, was a crucial component of the disaster that struck. Estimates of the number of deaths in the famine range from 15 million to 30 million, and the fact that we *don't even know* whether or not 15 million human beings died just adds an extra layer of horror to it.

You'd hope that the basic lesson of this—don't fuck with nature unless you're very, very certain what the consequences will be, and even then it's probably still not a good idea—would have stuck. But that seems unlikely. In 2004, the Chinese government ordered the mass extermination of mammals from civet cats to badgers in response to the outbreak of the SARS virus, suggesting that humans' capacity for learning from their mistakes remains as tenuous as ever.

Don't Underestimate Birds, Part II: Shakespeare in the Park

Eugene Schieffelin made basically the same mistake as Chairman Mao, except in the opposite direction. And where Mao's error was driven by a combination of public health goals and dictatorial fiat, the havoc that Schieffelin caused in his ecosystem—a man-made natural disaster that continues to this day—was driven entirely by whimsy.

What Schieffelin did one cold early spring day in 1890 has ended up spreading disease, destroying hundreds of millions of dollars' worth of crops every year and even killed 62 people in a plane crash. Which is quite a lot of damage for someone who was just trying to show off what a huge fan of Shakespeare he was.

Schieffelin was a well-to-do drug manufacturer who lived in New York City, but despite the strong potential for damaging screw-ups in that line of work, his contribution to environmental chaos stems not from his profession, but rather from his hobbies. He was extremely keen on two fashionable trends of the age—an absolute devotion to the works of Shakespeare, and transplanting species into new habitats.

At the time, Western culture was going through an all-consuming Shakespeare revival, with the result that the Bard had attained a status in popular culture that was at roughly Beyoncé levels. Meanwhile, based on a French idea, groups called "acclimatization societies" had started spreading around the Western world—voluntary groups of wealthy do-gooders who devoted themselves to introducing foreign species of plants and animals to their countries. (This was many years before people would twig just what an awful idea that could be.)

Schieffelin's mistake stemmed from the fact that he was the chairman of the American Acclimatization Society, based in New York, and also that he absolutely bloody loved Shakespeare. And so he hit on a delightful, eccentric plan: what better way to honor the greatest poet in the English language, he thought, than to introduce every single species of bird mentioned in Shakespeare's plays to the USA? The American Acclimatization Society set to work.

At first, they ran up against a string of failures: birds such as skylarks, bullfinches and song thrushes were released into the wild (well, the city, at least) but failed to take hold, dying out after a few years in the unfamiliar environment. But then, on March 6, 1890, Eugene Schieffelin stood in Central Park with his assistants and began opening a number of cages that contained a total of 60 European starlings.

You can't really blame Shakespeare for all this, but if he'd chosen a different bit of hyperbole in Act I, Scene III of *Henry IV, Part I*, then things would have been very different. In that scene, the character Hotspur, describing his determination to keep pressure on the king to pay a ransom for his brother-in-law Mortimer (despite the king forbidding him to even mention Mortimer's name), says:

> *Nay,*
> *I'll have a starling shall be taught to speak*
> *Nothing but "Mortimer," and give it him*
> *To keep his anger still in motion.*

That's the only time Shakespeare ever mentions starlings. The whole of the rest of the complete works, not a dicky bird. But that single reference was enough for our Eugene.

Those initial 60 starlings were released in 1890, and in 1891 Schieffelin went back and released 40 more. Initially, it didn't look great for the first American starlings—within a few years of bitter New York winters, only 32 of the original hundred were still alive, and it looked like they might follow in the wingbeats of their unlucky predecessors. But starlings are tough, versatile creatures, adept at fitting into new environments and bullying their way to survival. In an impressive bit of irony, a small flock of them found shelter from the elements under the eaves of the American Museum of Natural History—a building dedicated to preserving the nation's natural history inadvertently helped to alter that history dramatically. Because gradually, the starlings' numbers began to grow. And grow. And grow.

Before the decade was out, they were common across New York City. By the 1920s, they'd spread halfway across the country. By the 1950s, they were in California. Today, there are 200 million of the buggers living all across North America, and you can find them everywhere from Mexico to Alaska.

They have become, in the words of the *New York Times*, "one of the costliest and most noxious birds on our continent"—or, as the *Washington Post* once described them, "arguably the most hated bird in North America." Flocking together in huge murmurations that can number up to a million birds, they destroy crops on a vast scale, tearing through wheat fields and potato fields alike and obliterating grain stores. They are aggressive, chasing native bird species out of their nests, and they help spread diseases that affect both humans and livestock, from fungal infections to salmonella. They shit absolutely everywhere, and it smells awful.

Their massive flocks also pose a danger to air travel—in Boston in 1960, an estimated 10,000 starlings flew into a plane as it took off from Logan Airport, destroying its engines and sending it crashing to the ground, where 62 of the 72 passengers on board died.

Starlings are a pest, a health hazard and a significant financial drain on the agricultural economy of North America. The only reason they're even present on the continent is because a nice upper-middle-class chap was way too into his hobbies and didn't stop to think about the potential consequences. If he'd got into jogging or home-brewing or watercolors instead, none of this would have happened.

On the plus side, I guess, they probably help keep the insect population down?

5 MORE SPECIES WE PUT IN PLACES THEY SHOULDN'T BE

Cats

Everybody loves cats. Except in New Zealand, which didn't have any predatory mammals until we brought them with us—which was bad news for the local species, particularly the plump, flightless parrot the kakapo.

Cane Toads

Like the rabbits, cane toads (natives of South America) were introduced into Australia with good intentions—in this case, to eat a pest, cane beetles. They didn't eat the cane beetles. They ate almost everything else, though.

Gray Squirrels

When the American gray squirrel was introduced to Britain and Ireland, it immediately started throwing its weight around and bullying the native red squirrel close to extinction.

Asian Tiger Mosquito

A particularly annoying and potentially disease-spreading mosquito (it feeds at all hours, unlike many other species), it's notable for how it hopped continents—it traveled from Japan to America in 1985 in a shipment of used tires.

Northern Snakehead

Look, if you're going to introduce an Asian species into America, maybe don't make it a ravenous carnivorous fish that can walk across land and survive for days out of water? That's just asking for trouble.

4

FOLLOW THE LEADER

As human societies grew more complex, with villages becoming towns becoming cities, we were forced to confront a problem that's common to any large group faced with a complicated task—whether that's founding a civilization or working out where to go for dinner. Ultimately, you need someone to make a decision.

We don't know much about how the earliest human societies organized themselves. Human nature being what it is, it's a good bet that there have always been people who liked bossing other people around, but it's not entirely clear when this became an actual job rather than just a hobby.

What we do know is that (as already mentioned) not long after the origin of agriculture, humanity invented inequality. Well done, humans. Archaeologists can tell this by looking at the sizes of houses in early settlements. To begin with, there's not much difference between them. The societies seem to be fairly egalitarian. But over the first few thousand years after humans began planting crops, an elite starts to emerge who have much larger and fancier houses than everybody else. In the Americas, this rising inequality seems to hit a plateau after about 2,500 years of agriculture; but in the Old World, it just keeps on going up and up. Why? One possible explanation is that the Old World had draft animals like horses and cattle, which could be used for transport and to plow fields, which better enabled the creation of personal wealth that could be passed on down the generations. And thus the 1 percent were born.

At some point, these elites stop just being a bit richer than everybody else, and start actually ruling over them. Spiritual or religious leaders were probably the closest thing the earliest city-states had to rulers, but then something changes around 5,000 years ago in both Egypt and Sumer (modern-day Iraq), and we get the first examples of everybody's favorite mode of government—absolute dynastic monarchy! There's a Sumerian stone tablet that very helpfully lists all the kings (and a single solitary queen) in order, which means that it's possibly a record of the first kings in human history. Unhelpfully, however, a lot of it is clearly bollocks. The first king mentioned on it, Alulim, is recorded as having ruled for 28,800 years, which frankly seems unlikely given that it'd mean he'd still have over 22,000 years of his reign left today.

Why, exactly, does humanity opt again and again for the "put one dude in charge of everything" approach to decision-making? Obviously, they may not have had much choice: the first rulers might have seized power by force, or some other form of coercion. But it also seems likely to be linked to war—the pharaonic dynasty in Egypt begins when Egypt is unified by conquest, and the kings of Sumer emerge during a period of growing intercity conflict. A little while later, in 2334 BCE, after a few hundred years of Sumerian kings, they were conquered by neighboring king Sargon of Akkad, who was busy establishing the world's first empire. In Mexico, in the valley of Oaxaca, archaeologists can see this all play out in one location. The settlement of San José Mogote starts out as a small, egalitarian, nonhierarchical village shortly after the adoption of agriculture around 3,600 years ago. Over the next millennium or so, minor conflicts with neighboring villages escalate, and wealth and inequality increase, until by 2,400 years ago it's being ruled as a chiefdom, the valley is falling into a state of war and the population of San José Mogote has moved up a mountain and started building a defensive wall.

"Which came first, the leaders or the wars?" is a bit of a chicken-and-egg question, but the two certainly seem to go hand in hand—and unfortunately for everybody else, it's not something you can really opt out of if you'd rather stay as a small, egalitarian village. In good news for fans of war, we'll have more on war in a couple of chapters, but for now let's focus on the leaders.

I know it's hard to believe in our fortunate, enlightened times, but occasionally the people who become the leaders of countries aren't actually terribly well suited to the job. In fact, it's actually not that surprising: you probably have to be at least a bit weird to start with to even want to run a country. Some of us have enough trouble choosing what socks to wear in the morning—imagine actually wanting to choose what socks a whole nation should wear?

Of course, there are lots of different types of leaders, and plenty of ways that countries can end up stuck with them. You've got your different flavors of autocrat: the hereditary dynasties, the ruling by divine right, the seizing of power by force and various types of dictator. Oh, and you've also got democratic elections. We'll take a quick spin through the screw-ups of democracy in the next chapter; in this chapter, we'll take a look at some of history's most incompetent, awful and just plain weird autocrats.

Let's start with Qin Shi Huang, the first emperor of China, a man who shaped our modern world to a quite staggering degree through his combination of a farsighted vision and a brutal but effective approach to getting things done. Unfortunately for him, he also blew it with a classic bit of delusional supervillain-style overreach.

Qin united the seven warring kingdoms of China into a single country through the cunning diplomatic tactic of conquering all of them. Nobody had ever managed this before: in 222 BCE, at a time when Rome was only just beginning to ponder properly expanding beyond Italy and getting itself an empire, Qin was founding a vast political entity that would outlast them all.

Not only did he manage that, but he did so while instituting a series of reforms that would set standards for how a modern country should be organized: reducing the influence of feudal lords and establishing a centralized bureaucracy, standardizing writing, money and measurement systems and building key communications infrastructures such as a huge network of roads and an early mail service. Oh, and he started work constructing the first sections of what would become the Great Wall.

So...what's so wrong with Qin, then? Well, unfortunately, he did all this by suppressing all opposition, banning opposing philosophies, executing people who disagreed with him and violently forcing peasants into slavery for his construction projects. That probably doesn't come as much of a surprise, given how things in history tend to play out.

What is a bit more surprising is what he used all his unprecedented centralized power and widespread communication networks for: in short, to make his subjects hunt for the elixir of life.

Qin, being an ambitious sort, was obsessed with immortality, and was convinced that by unleashing the power of his new state he could brute-force his way to finding the secret of everlasting life. He sent out demands across the nation, roping in everybody from doctors to soldiers to tradesmen in the most remote parts of the country to contribute to his search. He ran his personal quest like a major government initiative, with his central court receiving regular progress reports from various outposts, and samples of herbs and potions being sent for his consideration. As part of it, all doctors had to register with the state. In some ways, it was an early form of centralized health system. In other ways, it really wasn't.

Sadly for Qin, it didn't work out as a great health system from his point of view. In true supervillain form, his search for immortality was his downfall: it's believed that many of the would-be elixirs of life he sampled contained mercury. Which, naturally, killed him. (And quite probably drove him mad from mercury poisoning before he died, which obviously is exactly what you want in a power-hungry absolutist ruler whose every word is law.)

By the time he died, everybody was so pissed off with Qin that they revolted almost immediately after he left the stage, overthrowing his heir a few years after his death. Qin's dynasty didn't last, even if the country he founded is a superpower to this day. They never did find the secret of everlasting life, though.

Sticking with China, but fast-forwarding about 17 centuries to 1505, if you want a helpful guide on why it's best not to put someone with the temperament of a spoiled child in charge of a country, then the Zhengde Emperor (born Zhu Houzhao) is probably a pretty good place to start.

His distaste for actually doing any of the work of ruling, when he'd much rather be off hunting tigers or sleeping with absurd numbers of women, was one thing. Not ideal, but, eh, you work with what you've got.

What was weirder was when he invented an alter ego for

himself—a dashing military leader called General Zhu Shou—and started giving this imaginary general orders to go and fight battles in the north, which, in character as Zhu Shou, he would of course dutifully follow. And which, by a remarkable coincidence, just happened to take him away from his work for many months.

That was definitely weird.

But probably not quite as weird as the fact that he had a full-size replica of a city market built inside the palace grounds, and would force all his most senior officials and military leaders to dress up as shopkeepers and play-act at being tradesmen so that he could dress like a commoner and walk around the market pretending to buy things. And if he caught any of them looking even a bit grumpy about this profoundly humiliating waste of time, they'd be fired, or worse.

Yep, that bit was probably the weirdest.

Oh, and there was also that time he decided it was a good idea to store all his gunpowder inside the palace just before a lantern festival. Which ended pretty much exactly how you would expect it to: explosively. (He survived the fire, but then died at the age of 29 from a disease he caught falling out of a boat. Twat.)

One problem with hereditary systems is that they do quite often end up with someone being in charge who clearly would rather be doing anything else but ruling. It was the case with the Zhengde Emperor, and it was also the case with poor Ludwig II of Bavaria. Unlike most of the other rulers on this list, "Mad King Ludwig" was mostly harmless; he just wasn't remotely into any of the things that were expected of the King of Bavaria. Instead, he preferred to devote his life to making things extremely fabulous.

When you look at the history of supposed madness in rulers, it's hard not to spot that many cases that feature on lists of "maddest monarchs" have something in common. Namely, the people in charge of writing history seem to be using "insanity" or "eccentricity" as a code for "insufficiently heterosexual." (Shout-out in particular to Queen Christina of Sweden, who refused to marry, preferred wearing masculine clothing and having uncombed hair and had what today would probably be referred to as a "gal pal." When put under pressure to find a husband, she instead renounced the throne, left Sweden dressed as a man and moved to Rome, where she entered the city on horseback dressed as an Amazon.)

We can only ever tentatively guess at the actual sexual orientation of historical figures (and we need to remember that the idea of "gay" as a specific, distinct identity only became solidified in Western societies within the last 150 years or so). That said, it still seems a pretty safe call to state that Ludwig II was super-super-gay.

Ludwig was a shy, creative daydreamer who was profoundly uninterested in the business of politics or leading an army. Instead, when he became king in 1864 at the relatively tender age of 19, he withdrew from public life and dedicated his reign to becoming a patron of the arts. What's more, he was pretty good at it.

He poured resources into the theater, hiring top talent and turning Munich into a cultural capital of Europe. He was a devoted fan of Wagner and became his personal patron, funding and supporting the composer to produce his late-career masterpieces after everybody else tried to run him out of town for being a knob. And above all, there were the castles.

Ludwig wanted Bavaria to be filled with fairy-tale castles. Getting theatrical stage designers rather than architects to plan them, he spent extravagantly on a series of increasingly ornate and flamboyant palaces—Schloss Linderhof, Herrenchiemsee and particularly the dramatic Schloss Neuschwanstein, perched on a rocky Alpine outcrop near his childhood home.

Schloss Neuschwanstein

All of this was very troubling to the great and the good of Bavaria. It wasn't exactly that Ludwig was inattentive to his duties—he would speed through his paperwork so he could get back to his true passions—but he was piling up debt to fund his artistic endeavors, hated appearing at public functions and his main interest in military matters seems to have been that the cavalry was full of hot guys.

And then there was the issue of an heir. As kings usually were, Ludwig was under constant pressure to marry and have children. He got engaged to a duchess who shared his love of Wagner, but as the wedding date came closer, he postponed it over and over again, before finally calling it off. He never even came close to marrying again.

Eventually, as Ludwig's debts increased and his plans for future castles got more and more elaborate, his enemies at court decided to act, and followed the time-honored route of having him declared insane. Now, the idea that some mental health issues might have run in Ludwig's family isn't out of the question (his aunt Alexandra thought she had a glass piano inside her body, although that didn't stop her going on to have a literary career). But of the four eminent doctors that the conspirators persuaded to sign off on Ludwig's diagnosis, none had ever examined him, and only one of them had ever even met him (12 years earlier). Among their evidence of his clear unfitness to rule was the damning fact that he forbade a servant to put milk in his coffee.

But the ruse worked, and despite the best efforts of a friendly baroness who temporarily fought off the government commissioners with her umbrella, Ludwig was deposed, and taken to be imprisoned (sorry, "treated for his health") in a castle south of Munich. The suspicion that not everything was entirely aboveboard about all this is only increased by the fact that three days later, Ludwig and his doctor were both found dead in a shallow lake, in what can only be described as "mysterious circumstances."

But in some ways, Ludwig had the last laugh. All those castles that he spent so lavishly on? They're now globally famous—Schloss Neuschwanstein is the iconic representation of Bavaria around the world—and attract millions of visitors a year, all of which is pretty good news for the Bavarian economy. If the plotters hadn't stopped Ludwig's future plans by deposing him, who knows how much more the Bavarians might have now. The person who fucked up here wasn't poor daydreaming Ludwig. It was them.

Even if you've never heard of Schloss Neuschwanstein, you've still seen it a hundred times. Its romantic turrets and spires were the direct inspiration for the castles in Disney's *Cinderella* and *Sleeping Beauty*, which themselves became synonymous with the world's biggest entertainment company. Any time you see that shooting star sprinkling its fairy dust over the castle in Disney's logo, you're watching Ludwig's dream living on.

Ludwig was far from the only leader whose dreams and talents lay in a different direction to the business of ruling. His love of building castles was at least a vocation that sits in vaguely the right ballpark for a monarch. A less suitable career would be one, say, as an enthusiastic and indefatigable pickpocket.

Now, if the only notable thing that Farouk I of Egypt had done in his life was to pickpocket Winston Churchill's watch while taking part in a crucial meeting during World War II, then he might be remembered a bit differently. He would have gone down in history at worst as a mild eccentric; at best, as an absolute legend who was basically the King of Banter.

But Farouk did not stop there.

Despite being richer than any of us could ever dream, Farouk—the second and final adult King of Egypt—just bloody loved stealing things. He would steal things from the great and the good, and he would steal things from commoners. He had one of the most notorious pickpockets in Egypt released from prison just so that he could teach him how to steal things better. When the body of the recently deceased Shah of Iran was resting in Egypt on its way to Tehran, Farouk literally stole a jeweled sword and other valuables from out of the coffin. (Unsurprisingly, this caused something of a diplomatic incident.)

It wasn't just stealing that marked Farouk out as perhaps not great king material. He was renowned for his appetite, partying and lavish lifestyle. Once described as "a stomach with a head," after becoming king as a handsome teenager he rapidly ballooned in size to over 280 pounds. He was so fond of his official car, a red Bentley, that he decreed that nobody else in Egypt could own a red car. He built up a vast collection of low-grade pornography. An inveterate and profligate gambler, he surrounded himself with a coterie of chancers, con artists and corrupt officials. Once, after having a nightmare in which he was being attacked by lions, he awoke and demanded to be taken to the Cairo zoo, where he promptly shot their lions.

He might have got away with all this if he hadn't been busy alienating people in other ways. The British had grudgingly recognized Egyptian independence in 1922, but still maintained

Farouk I of Egypt (1920–1965)

a large, unpopular military presence in the country, and many of Farouk's subjects increasingly saw the monarchy as a puppet of the West. For their part, the British were growing increasingly narked at Farouk for not being *enough* of a puppet. (For more on this sort of thing, see the later chapter on colonialism.)

When World War II came along, it wasn't just stuff like nicking Churchill's watch that turned everybody against Farouk. It was other little things, like refusing to turn off the lights in his Alexandria palace while the city was on blackout due to German bombing. Or sending Adolf Hitler a note saying that he would welcome a Nazi invasion, on the grounds that it might get rid of the British.

Farouk just about made it through the hostilities, belatedly declaring war on the Axis powers at roughly the point that the fighting was over, but didn't last long after that. He was deposed in a military coup in 1952 (his six-month-old son technically became king for just under a year before the monarchy was abolished) and lived out his remaining years in Monaco and Italy, where, as *Time* magazine wrote, he "grew ever more gross and more persistent in the pursuit of women." He eventually died in the time-honored manner of exiled leaders—of a heart attack, at the age of 45, during the cigar course after a massive dinner in a restaurant in Rome.

(For the record, Churchill did not find the watch thing funny, and angrily asked for it back.)

You'd hope that the quality of rulers we get might have improved a little over time, but there are plenty of leaders from the modern era who can rival their historical counterparts for baffling awfulness. For example, Saparmurat Niyazov, who ruled Turkmenistan for over 20 years, from when it was still part of the Soviet Union, through independence, until his death in 2006, stands as a prime example of the fact that you can always build a cult of personality around a dictator, even if that dictator's personality is extremely stupid.

For two decades, president-for-life Niyazov ruled the country according to his personal whims, almost all of which were deeply weird. He insisted on being referred to as "Türkmenbaşy," meaning "leader of the Turkmen." He banned dogs from the capital city of Ashgabat because he didn't like the way they smelled. He outlawed beards, long hair on men and gold teeth. He was keen on passing judgment on television personalities, and prohibited TV newsreaders from wearing makeup because he said it made it hard to tell the men and women apart. He banned opera and ballet and circuses, lip-synching at gigs, playing recorded music at events like weddings and even listening to the radio in the car.

He built a giant gold statue of himself in Ashgabat that rotated so that it always faced the sun. He absolutely loved putting his name on things. In 2002, he renamed the month of January "Türkmenbaşy," while April became "Gurbansoltan" after his mother. A major city was rechristened "Türkmenbaşy."

Gold statue of Saparmurat Niyazov (also known as "Türkmenbaşy") in Ashgabat

Bread was renamed after his mum. The airport in Ashgabat was named "Saparmurat Türkmenbaşy International Airport." He instituted a public holiday in honor of melons, specifically a new variety of muskmelon that was named, in a shocking twist, "Türkmenbaşy."

He wrote a book called the *Ruhnama*, which was part poetry collection, part autobiography, part dodgy history lesson and part self-help tract. Not liking the book was punishable by torture. Knowledge of the book was a required part of the state driving test. He closed down all libraries outside the capital city, on the grounds that the Quran and the *Ruhnama* were the only books anybody needed to read. He built a giant statue of his own book in the capital city, which rotated and played audio passages at regular intervals. Reading the book was declared to be a prerequisite for entry to heaven. (It was possibly ghostwritten.)

He spent vast sums on ridiculous buildings, like an ice palace in the desert, a giant pyramid and a $100 million mosque that he named "Spirit of Türkmenbaşy." He built a giant concrete staircase on a desolate mountain and forced every public servant to go on a 23-mile walk along it every year. In 2004, he sacked 15,000 medical staff from the country's health service and replaced them with soldiers; he closed all hospitals outside the capital, on the grounds that if people were sick, they could travel in; he swapped out the Hippocratic oath for an oath sworn to Türkmenbaşy. He reportedly used to seize smuggled shipments of drugs and keep them for himself, shooting pistols at imaginary enemies in his darkened residence. There was no free press, dissidents were suppressed and all public groups, political parties and religions had to register with the "Ministry of Fairness." Outside the Ministry of Fairness stood a giant statue of the figure of Justice—who, people couldn't help but notice, looked surprisingly like Türkmenbaşy's mother.

It's not entirely clear what broader lessons we can draw from Niyazov's long, extremely awful reign, other than if you ever catch yourself acting even a little bit like him, please, please stop.

But as bad as Türkmenbaşy was, and as unlucky as Turkmenistan was to suffer under the two decades of his reign, he still doesn't quite make it to the top of the "extremely regrettable autocrats" list. There have been leaders more evil, and possibly even leaders more incompetent. But if you want a good example of just how fucked-up autocracy can get, then it's hard to beat the period of the Ottoman Empire that proved bad things sometimes really do come in threes.

The Gilded Cage

Very few places have had a run of really terrible leaders quite like the one the Ottoman Empire suffered in the first half of the seventeenth century. Two of them usually have the words "the Mad" retrospectively added to their names, which is never a good sign. Worse, the one who doesn't even get called "the Mad" might have deserved it the most.

Given that two of them were brothers and the other was their uncle, it's hard not to suspect that something hereditary might have been going on there. But equally, there's also an overwhelming sense of "well, what did you *expect*?" If you were actually trying to set up a system designed to produce somewhat unstable rulers, it's hard to see how you could have done better than this.

The Topkapı Palace in Istanbul was not an especially safe place to be during this period, particularly if you were the son of the current sultan. The problem was your brothers—or at least, they were the problem as soon as the current sultan died and all of you tried to claim the throne at the same time.

As tended to happen with monarchies at the time, extremely bloody fights over the succession had effectively become a tradition in the previous centuries—a tradition that had an inconvenient habit of spilling over into prolonged civil war. This wasn't terribly convenient for anybody, especially when you had an empire to expand, so sons of the sultan usually decided it was more efficient to forestall any sibling rivalry by...well, by having all their brothers murdered.

The downside of this institutional fratricide was that the Ottoman dynasty was permanently vulnerable to coming to an abrupt end, if a sultan were to die without having any sons to take over and with no brothers left unmurdered. There was also the slight issue of Sultan Mehmed III, who had no fewer than 19 younger brothers murdered when he acceded to the throne in 1595, which everybody seems to have agreed was a bit much. So, starting with Mehmed III's successor, Ahmed I, a compromise was put in place: Kafes, which literally means "the Cage" and was a place to keep your spare brothers.

The Cage was not, in fact, a cage—it was a relatively luxurious, tastefully decorated tower just next to the harem—but it certainly had a few features in common with a cage. Like, for example, not being able to leave it.

When Ahmed I became sultan in 1603, he unexpectedly broke with the brother-murdering tradition and allowed his younger brother Mustafa to live. The fact that Ahmed was just 13 years old and Mustafa was 12 at the time might have played a part in this decision—Ahmed wouldn't even father a son until the following year. And partly it might have been due to him feeling sympathy with Mustafa, who seems to have already been quite fragile. Basically, there seems to be a possibility that Ahmed might have been...almost nice?

So anyway, rather than being killed, Mustafa was sent off to live in the Cage, while Ahmed I got on with being sultan. This all went swimmingly until 1617, when Ahmed died of typhus.

By this point, he had fathered a bunch of sons, who technically should have inherited the throne. But due to a combination of the fact that they were also pretty young, and various bits of palace intrigue (largely down to Ahmed's favorite consort, Kösem, not wanting her sons to be murdered when their older half brother came to power), the powers behind the throne decided to change the line of succession. Rather than going to Ahmed's eldest son, Osman, it would pass brother-to-brother. And so it was that Mustafa became Mustafa I.

It's fair to say that this did not go well.

Mustafa was really not cut out to be sultan. He doesn't seem to have been terribly keen on the idea, and the whole situation hadn't been helped by having spent the first 12 years of his life convinced that his brother was going to murder him, and then the next 14 years imprisoned with nothing to do but take opium and hang out with concubines. The powerful court eunuchs had hoped that being reintroduced into society might focus his mind a bit. Nope.

Mustafa's main approach to governing seems to have involved giggling an awful lot, pulling the beards of his viziers and knocking their turbans off while they tried to tell him important government things. He had a tendency to appoint random people—like a farmer he met while out hunting—to powerful official positions. He was also noted for being accompanied around the palace by two virtually naked slave women, and for his habit of trying to feed gold and silver coins to fish.

The whole thing lasted about three months before everybody had absolutely had enough, and Mustafa I was overthrown by the 14-year-old Osman. Somehow, for a second time, he managed to avoid being murdered, and instead got sent back to the Cage.

That would have been that, except for the fact that the precocious Osman II was an ambitious, unorthodox sultan with a zeal for reform, who refused to be bound by tradition. (Well, mostly. He did manage to squeeze in murdering at least one of his brothers during his reign, for old times' sake.) Osman made the crucial mistake of really annoying the elite units of the Ottoman army, the Janissaries—blaming them for a failure to win a battle he'd led, punishing them by closing their coffee shops and banning them from smoking or drinking, before finally planning to disband them altogether and raise an alternative army in Syria.

While Osman might have genuinely had a point about their military effectiveness, the Janissaries were, unsurprisingly, not entirely on board with this plan. So Osman II was given the distinction of becoming the first example of regicide in Ottoman history, killed by his own army through the inventive combination of strangulation and "compression of the testicles."

And then, in the absence of anybody else to take over, once again Mustafa was coming out of his Cage. And he was doing… not fine.

It's not clear if everybody thought that four more years' confinement might somehow have improved his mental state, but if so, they were quickly disappointed, as straight away Mustafa was extremely back on his bullshit. For starters, when they came to get him out of the Cage and told him he was sultan again, he barricaded himself inside and refused to come out, explaining (not unreasonably), "I do not want to be the sultan." After they managed to winch him out through a hole in the roof, he spent large amounts of his time running through the palace desperately looking for Osman II, who he believed was still alive and might be hiding in a cupboard. If he could find Osman, his reasoning went, *he* could take over being sultan again and Mustafa wouldn't have to do it anymore.

This all went on for another 17 months (during which period Mustafa did at least find time to put a donkey driver he'd met in charge of a major mosque) before everybody decided that enough was enough. Even Mustafa's mother signed up to the idea of deposing him for a second time, with the proviso that she would prefer it if they could see their way to not murdering him. Remarkably, everybody agreed, and Mustafa was sent off to live out his days in the Cage, having somehow managed to be sultan two times and murdered zero times.

The new sultan, Murad IV, had two major benefits for the power players in the Ottoman court: (a) he was not obviously mad, and (b) he was an 11-year-old child. His mother, Kösem, who was a remarkably skilled power player herself, got a good few years of ruling on behalf of a puppet sultan out of this arrangement. That was before Murad IV grew old enough to reveal that, if not actually mentally unsound, he was at least an utter, utter bastard.

Inheriting an empire that was in something of a state, he made up his mind to assert his authority. Hard. Deciding that his half brother Osman hadn't gone far enough with banning stuff just for the army, Murad banned smoking, drinking and especially coffee for everybody in the Ottoman Empire.

In a list of "moves designed to piss lots of people off," banning coffee in Turkey probably ranks somewhere alongside banning cheese in France, banning guns in America and...well, banning national stereotyping in Britain. But Murad was determined. He hated coffee drinkers so much that he would patrol the streets at night dressed in civilian clothes, looking for people drinking coffee and then executing them on the spot.

When not enforcing his strict anticoffee laws, he liked to wind down by executing people for literally any other reason he could think of: for playing the wrong kind of music, for talking too loudly, for walking or sailing too close to his palace or just for being women. *Especially* for being women. He really hated women.

By the end of his reign, Murad wasn't even really executing people anymore, as that implies he at least had some sort of vague pretext. He was pretty much just running around with a sword, pissed out of his skull, killing any poor bastard he came across. Some estimates suggest that he might have personally executed around 25,000 people during just 5 years of his 17-year reign—which on average would be more than 13 every single day. Again, it's really worth emphasizing that this is the guy who *doesn't* get "the Mad" attached to his name.

Oh, and obviously he also murdered most of the rest of those brothers that Osman had left not-murdered.

When Murad IV died in 1640 (of cirrhosis of the liver, which must have come as a bit of a surprise to his subjects, who he'd banned from drinking alcohol), there was in fact just one non-murdered brother remaining: Ibrahim. By this point Ibrahim had spent virtually all of his 25 years of life confined to the Cage, living in perpetual fear of his seemingly inevitable murder. He wasn't entirely wrong about that: Murad did in fact order Ibrahim's murder from his deathbed, preferring to see the Ottoman dynasty die out entirely than Ibrahim come to the throne. The only reason the murder didn't happen was that, as is often the case with bickering brothers, their mother, Kösem, stepped in and stopped it.

But if everybody was tempted to breathe a sigh of relief now that Murad was out of the picture, Ibrahim soon cured them of that mistake. Because if he hadn't been insane when he went into the Cage, he certainly was when he came out.

Much like Mustafa before him, he was initially reluctant to

come out of the Cage at all, as he was convinced it was a huge trick on the part of Murad, all the better to murder him with. The only thing that would reassure him was when they actually brought him Murad's dead body.

Once they'd coaxed him out, Kösem—perhaps realizing that he wasn't terribly well suited to ruling—suggested that he might like to busy himself with some concubines instead. Unfortunately, Ibrahim took her suggestion to extremes.

In addition to his other proclivities (like being obsessed with fur, wearing fur coats all the time and demanding that every room in his palace was decked out with huge quantities of fur), Ibrahim was sexually obsessed and virtually insatiable. This suited Kösem, who was busy ruling in his place—she had Ibrahim supplied with a vast number of slave girls and kept him hopped up on aphrodisiacs so that exhaustion and impotence wouldn't leave him sexually incapable for long enough that he might accidentally do some ruling of his own.

Ibrahim's sexual habits included—I'm going to be honest here—some extremely grim shit. As a prince of Moldavia, Dimitrie Cantemir, wrote some years later: "In the palace gardens he frequently assembled all the virgins, made them strip themselves naked and, neighing like a stallion, ran among them and, as it were, ravished one or the other, kicking or struggling by his order."

It gets worse. According to Cantemir, one day Ibrahim saw a wild cow when he was on a trip, and became obsessed by its genitals. So much so that he had a cast made of them, and then copies of the cast crafted in gold and sent all around the empire, with an order that his servants find a woman whose genitals could match the cow's.

Yeah.

(One caveat: it's worth noting that Cantemir might not be a *wholly* unbiased source. On the one hand, he had lived and studied in Constantinople and spoke Turkish, and was writing only a few decades after the events. On the other hand, his book was called *The History of the Growth and Decay of the Ottoman Empire*, and he wrote it shortly after switching Moldavia's allegiance from the Ottomans to Russia, losing catastrophically in battle, then being deposed and exiled, so he might have had a slight grudge. The supposedly "decaying" Ottoman Empire lasted in some form for another two centuries.)

Ibrahim's search for his ideal woman, whether or not prompted by a bovine encounter, ended with her being found in Armenia. She was named Sugar Cube, and she quickly became Ibrahim's favorite. Things start spiraling out of control: Sugar Cube told Ibrahim that one of his other concubines had been unfaithful, which sent Ibrahim into such a rage that he slashed his own son's face with a knife for joking about it, and then—being unable to tell which woman was the supposedly guilty one—had all but two of his 280-strong harem tied up in sacks and drowned in the Bosphorus. Only one survived. Sometime after this, fearing Sugar Cube's growing influence, Kösem invited her over for dinner and a little girl chat, during which she quickly murdered her. (She told Ibrahim that Sugar Cube had died of a sudden illness.)

By this point, Ibrahim's excesses had alienated pretty much everybody, and the cost of keeping him in his lavish lifestyle of sex and fur was draining public funds. He had several sons, and so the dynasty was no longer under threat. Even Kösem agreed things had gone too far, and signed off on a plan to depose him. And so for the second time in a couple of decades, the Janissaries revolted; a mob dismembered the grand vizier, and they marched Ibrahim back to his dreaded Cage. He spent the last 10 miserable days of his life back in the same place where he'd spent most of his childhood, before the plotters decided to take the quick route out and murdered him.

The history of this period in the Ottoman Empire reads so much like a bloody, misogynist fever dream—something that makes *Game of Thrones* look like an episode of *The Joy of Paint-*

ing with Bob Ross—that it's occasionally hard to believe. And to be sure, it's another case where it's sometimes difficult to distinguish what was real and what was simply propaganda to justify all the political upheaval and murders.

The story of this time in history isn't just one of crazed men, and a few powerful women trying to keep things stable; across large parts of the world it was an era of new technology and dramatic economic shifts, with allegiances in flux, borders being redrawn and wars all over the place. The Ottoman Empire was no exception. By the second half of the seventeenth century, when they finally left this period of instability, the Ottomans had waved goodbye to the era of institutional fratricide and civil war, had a newly monetized economy and had effectively changed their system of government from a feudal absolutist monarchy to a modern bureaucracy. So far from this being the point that marked the start of the Ottoman Empire's decline, on the whole, they actually came out of it all pretty well!

That's probably not much comfort to all the people who got murdered, though.

5 MORE LEADERS WHO REALLY SHOULDN'T HAVE BEEN PUT IN CHARGE OF ANYTHING

Kaiser Wilhelm II

Germany's Wilhelm II believed himself to be a master negotiator with a diplomatic golden touch. In fact, his only gift was insulting just about every other country he came into contact with, which may help explain how World War I happened.

James VI and I

Not the worst king ever—he unified the crowns of Scotland, England and Ireland and commissioned a solid Bible—but he was obsessed with witch hunting, personally supervising witch torture and writing a book about his great witch-hunting exploits.

Christian VII

Christian VII of Denmark was a poor king in many ways, but probably his obsessive, uncontrollable masturbation was the least kingly aspect of it.

Tsar Peter III

Was just really into toy soldiers. Didn't consummate his marriage to Catherine (later "the Great," after she deposed him) for years because he was too busy playing with them, and once had a rat court-martialed after he found it nibbling one of his toys.

Charles VI

Best known for the delusional belief that he was made of glass and might shatter at any moment, Charles VI of France's sad reign ended shortly after the English tricked him into signing a treaty that declared the English monarchy heirs to the French throne—basically guaranteeing many more years of war.

5

PEOPLE POWER

Thanks to the capacity of autocratic rulers to make grand, operatic screw-ups on a horrifying scale, over the course of history various states have tried to mitigate this by trying out a little thing called "democracy." With, it has to be said, varying degrees of success.

Quite where democracy was first tried out is somewhat disputed—forms of collective decision-making were almost certainly a feature of early small societies. There's also some evidence for something at least approaching democracy in India about 2,500 years ago. But generally, it's the Greek city-state of Athens that gets the credit for adopting and codifying democratic government at around the same time, in 508 BCE.

Of course, many of the key features of a democracy (government being open to all citizens, and elections in which citizens can replace their government if they don't like it) do rather depend on who gets to count as a citizen. And for much of history, across many countries, that hasn't included a few insignificant little categories of people—such as women, or poor people, or ethnic minorities. I mean, you can't give power to just *anybody*, right?

Another problem with democracy is that people are generally big fans of it when they think it might give them power, but suddenly become notably less keen when it looks like it might take power away from them. As a result, democracy often involves a frankly exhausting amount of work simply to ensure it keeps on existing.

For example, Rome experimented with various cunning techniques to stop democracy sliding into autocracy. One was to split the position of consul—the most powerful elected role, which combined both civil and military leadership—between two people. They'd be elected for a year, would swap holding the most significant powers every month and were each given command of two of the four legions of the Roman army. Which is a pretty clever way to make sure that absolute power doesn't fall into any single man's hands.

Unfortunately, it wasn't ideal when all four legions of the army were required for a single battle—as happened at the Battle of Cannae in 216 BCE, when Rome was faced with the assembled might of the Carthaginian forces commanded by noted elephant fan Hannibal. In that case, command of the army swapped between the two consuls, Lucius Aemilius Paullus and Gaius Terentius Varro, *on a daily basis*. A problem that was compounded by the fact that they didn't agree on tactics. One day the cautious Paullus would be in charge, the next the more reckless Varro, and so on. Hannibal, who wanted to draw the Romans into battle, simply waited a day until Varro was in charge, at which point he got his wish. The result was the Roman army was virtually wiped out.

The Romans actually had a way of stopping this kind of division happening—they would appoint a "dictator," one man who would be given absolute power in times of crisis, on the understanding that he would resign once the specific job he'd been given the power for was done. (Ironically, just before the Battle of Cannae, the Roman Senate had got rid of a dictator because they didn't like his tactics.) Again, this was a great idea in theory, but it did rather rely on the person you'd just gifted with absolute power and the command of a vast army subsequently giving it up willingly. Which they mostly did, until an ambitious chap called Julius Caesar decided that he quite liked the power and actually he'd rather keep it, if it's all the same to you. That ended stabbily for Caesar, but his successors also decided that absolute power was an excellent thing to have, and so the Roman Republic quickly turned into the Roman Empire.

Some of the approaches that democratic systems have taken to prevent the power-hungry from gaining undue influence over proceedings have been quite remarkable. If you get confused by, say, the electoral college system in the United States, then be thankful you didn't live in the Republic of Venice. Several centuries before the word *doge* became a popular internet dog meme featuring a picture of a baffled yet placid Shiba Inu, Venice was ruled by a doge, a leader who was elected by possibly the most complicated electoral college system ever.

Given that the doge was elected for life by a Great Council of a hundred or so oligarchs—a setup with obvious potential for corruption—the electoral system was established in 1268 with the intent of preventing anybody from being able to fix the election. Here is how the Doge of Venice was elected: first 30 members of the council were chosen by randomly drawing lots. From these, lots would be drawn again, reducing the number of electors to 9. Those 9 would then elect 40 council members, who would then be whittled down to 12 by lot. Those 12 would elect 25 members, who would again be reduced by lot to 9, who would elect 45, who would be reduced by lot to 11, who would then elect 41 members—and finally, on the tenth round of the whole process, *those* 41 would elect the doge.

Try reading that out loud without taking a breath.

This is obviously completely ridiculous, and must have been a bloody nightmare for Venetian political pundits trying to make predictions. But in fairness to the oligarchs of Venice, it does seem to have been pretty successful (if you were a Venetian oligarch, that is) as the system stayed in place for over 500 mostly prosperous years, until the Republic of Venice was finally conquered by Napoleon Bonaparte in 1797.

Frankly that makes Venice a beacon of stability, especially when you consider that at the time of writing, Italy has, notoriously, had 66 governments and 44 prime ministerships in the 73 years of the postwar period. By contrast, the UK has had just 15 prime ministerships over the same period (in both cases, some people have held the role more than once, hence "prime ministerships" rather than "prime ministers"). That "at the time of writing" is quite important, because Italy is currently governed by a coalition of populists and far-right nationalists that doesn't exactly scream stability. By the time it's published, they may be on to government number 67 and prime minister 45 or possibly more. As such, in the interests of accuracy, here is that fact again, with an empty space so you can write an updated number for how many governments Italy has had:

Italy has had [   ] governments since 1946.
(Please visit *howmanygovernmentshasitalyhad.com* for the current figure. And maybe write it in pencil?)

One of the problems with democracy's fragility is that policies that may have seemed reasonable under a nice fluffy liberal democracy can start to backfire rather horribly when a more authoritarian regime takes over. For an example, look to Mexico in the first half of the nineteenth century, when the Mexican authorities—newly independent from Spain—decided to put the underdeveloped land in their northern province of Texas to good use. Wanting a buffer zone that would protect Mexico from both raids by the Comanche people and the westward growth of the United States, the Mexicans started encouraging American ranchers and farmers to come and settle in the area, handing over large tracts of land to *empresarios*, agents who would encourage Americans to make the move (the fact that there was no extradition treaty may have been a big factor for some people).

They started to realize that this was going a bit wrong when it became apparent that some *empresarios* were gaining significant political power—and many settlers were unwilling to integrate and obey the laws of the Mexican government. Freaked out, in 1830 the Mexicans abruptly tried to ban any further American migration, but found themselves powerless to stop the influx of American immigrants pouring across the border.

Things came to a head when the (relatively) liberal Mexican government was replaced by an autocratic, authoritarian ruler in president Antonio López de Santa Anna, who in 1835 dissolved the Mexican congress and pushed through major changes to the country's constitution that centralized power, effectively making him a dictator. He also started to forcefully suppress dissent in Texas, a crackdown on the American immigrant community that only inflamed tensions further—and soon a full-scale rebellion was on the cards. By 1836, after a war that included the infamous events of the Alamo, Texas had declared its independence. By 1845, it was part of the ever-expanding United States, and rather than having a useful buffer against American expansion, Mexico had lost a valuable province.

There are a couple of divergent lessons we can draw from this. On the one hand, there's "don't encourage immigration only to later turn against those same immigrant communities."

On the other, there's also "don't assume that you'll always be a democracy because THAT'S EXACTLY WHEN THINGS GO WRONG."

Democracy, of course, does rely somewhat on the voters making good decisions in the first place. For example, in 1981 the small Californian town of Sunol elected a dog as their mayor. Bosco Ramos, a black Labrador mix, beat two human candidates in a landslide victory after his owner, Brad Leber, entered him into the race following an evening of talking shit at a local bar. In fairness to Bosco and the voters of Sunol, this actually seems to have worked out fine—Bosco was widely hailed as a very good boy, and served as mayor for over a decade, ending only with his death in 1994. One resident recalled to the *San Jose Mercury News* in 2013 that the mayor "used to hang in all the bars and he used to growl at you if you didn't feed him," and he was rumored to have fathered numerous puppies with different bitches around town, which sounds like pretty standard politician behavior to be honest. Bosco is fondly remembered in Sunol, where a bronze statue of him stands to this day, and his tenure only involved one major international incident—when in the wake of the Tiananmen Square massacre the Chinese newspaper *People's Daily* used the example of Bosco to attack Western democracy on the grounds that "there is no distinction between people and dogs." Bosco ended up joining a group of Chinese students on a pro-democracy protest outside the Chinese consulate in San Francisco.

Bosco's election may have been unexpected, but he doesn't even come close to being the weirdest nonhuman victor of an election. That honor probably goes to Pulvapies, a brand of foot powder that was elected mayor of the Ecuadorian town of Picoaza in 1967. Pulvapies wasn't even officially standing in the election, but its maker did run a joke marketing campaign across the country with the slogan "Vote for any candidate, but if you want well-being and hygiene, vote for Pulvapies." Come election day, Pulvapies received thousands of write-in votes in several areas—and in Picoaza, the foot powder somehow managed to come out in first place, much to the chagrin of the numerous human candidates.

Still, as unorthodox as electing nonhuman politicians might be, if you want to achieve a really impressive democratic screw-up, your best bet is still to elect a human—as demonstrated by the fact that making a brand of foot powder mayor isn't even the worst electoral decision in Ecuador's recent history.

Instead, that honor probably goes to the election of Abdalá Bucaram as the country's president in 1996. Bucaram, a former police commissioner, mayor and occasional rock singer who campaigned under the self-bestowed nickname "El Loco" ("The Madman"), was swept to a shock victory after a populist presidential campaign that attacked the country's elites. As a police commissioner he had been notorious for the way he "chased down women wearing miniskirts, jumping off his motor scooter and ripping out hems to make their skirts longer," as the *New York Times* reported when he was elected. As a mayor he also had a track record of shaking down local businesses for payments, and in 1990 he had fled to Panama to avoid corruption charges. During the presidential election campaign, his unconventional rallies and campaign adverts—which often involved him singing, accompanied by the band that went everywhere with him on the campaign trail—galvanized the country's working class, who were promised that Bucaram would bring an end to the neo-liberal policies of privatization and austerity that the nation's political class were committed to. Things that might have proved career-ending for other politicians—you know, stuff like the way he had a Hitler mustache and once said that *Mein Kampf* was his favorite book—didn't seem to be much of a barrier to his success.

Once he was in power, the country's poor who had voted for him were somewhat surprised at the economic plan he unveiled a few months into his term: a neo-liberal program that extended privatization and doubled down on austerity, the exact things he'd been elected to stop. Oh, and he tried to remove the term limits on the presidency. And went off-script in his speech announcing the economic policy to mount a lengthy attack on a newspaper that had been critical of him.

He continued his commitment to eccentric behavior while in office, including releasing a song titled "A Madman Who Loves," meeting with Lorena Bobbitt (the woman who became famous for cutting off her husband's penis) and selling his Hitler mustache for charity. Also, if press reports at the time were accurate (again, it can occasionally be hard to tell what accusations are true and what are just scuttlebutt), he also put his teenage son in unofficial charge of the customs service, and they reportedly threw a party celebrating the son making his first million dollars. The minimum wage in Ecuador at the time was $30 a month, so you can see why that might have annoyed some people.

Not unsurprisingly, popular opinion rather quickly turned against Bucaram, prompting massive street protests against his reign, and he was impeached and removed from the presidency a mere six months into his term, on the grounds that he was "mentally incompetent." (That was almost certainly just a pretext, but if you're going to campaign as "El Loco," then you probably haven't left yourself much of a leg to stand on.) He was also charged with embezzling millions of dollars, and promptly fled—again—to exile in Panama. There are various lessons that we can take from all this, but probably the main one is: "If somebody has a Hitler mustache, then, uh, that might be a bit of a red flag?"

Speaking of which...you can't really talk about democracy's capacity to go rapidly and nightmarishly wrong without talking about, well, Hitler.

Hitler

Look, I know what you're thinking. Putting Hitler in a book about the terrible mistakes we've made as a species isn't exactly the boldest move ever. "Oh wow, never heard of him, what a fascinating historical nugget" is something you're probably not saying right now.

But beyond him being (obviously) a genocidal maniac, there's an aspect to Hitler's rule that kind of gets missed in our standard view of him. Even if popular culture has long enjoyed turning him into an object of mockery, we still tend to believe that the Nazi machine was ruthlessly efficient, and that the great dictator spent most of his time...well, dictating things.

So it's worth remembering that Hitler was actually an incompetent, lazy egomaniac and his government was an absolute clown show.

In fact, this may even have helped his rise to power, as he was consistently underestimated by the German elite. Before he became chancellor, many of his opponents had dismissed him as a joke for his crude speeches and tacky rallies. He was a "pathetic dunderhead" according to one magazine editor; another wrote that his party was a "society of incompetents" and that people should not "overestimate the fairground party."

Even after elections had made the Nazis the largest party in the Reichstag, people still kept thinking that Hitler was an easy mark, a blustering idiot who could easily be controlled by smart people. Franz von Papen, the recently removed Chancellor of Germany, who was bitterly determined to reclaim power, thought that he could use Hitler as a pawn, and so entered into discussions with him to form a coalition government. After the deal was done in January 1933, making Hitler chancellor and von Papen vice chancellor, with a cabinet full of the latter's conservative allies, von Papen was confident of his triumph. "We've hired him," he reassured an acquaintance who tried to warn him he'd made a mistake. "In two months," he predicted to another friend, "we'll have pushed Hitler so far into a corner he'll squeak."

That's not how it worked out. In fact, within two months, Hitler had seized complete control of the German state, persuading the Reichstag to pass an act that gave him power to bypass the constitution, the presidency and the Reichstag itself. What had been a democracy was, suddenly, not a democracy anymore.

Why did the elites of Germany so consistently underestimate Hitler? Possibly because they weren't actually wrong in their assessment of his competency—they just failed to realize that this wasn't enough to stand in the way of his ambition. As it would turn out, Hitler was really bad at running a government. As his own press chief, Otto Dietrich, later wrote in his memoir *The Hitler I Knew*, "In the twelve years of his rule in Germany, Hitler produced the biggest confusion in government that has ever existed in a civilized state."

Hitler hated having to read paperwork, and would regularly make important decisions without even looking at the documents his aides had prepared for him. Rather than having policy discussions with his underlings, he'd subject them to impromptu rambling speeches about whatever was on his mind—which they dreaded, as it would mean no more work could be done until he was finished.

His government was constantly in chaos, with officials having no idea what he wanted them to do, and nobody was entirely clear who was actually in charge of what. He procrastinated wildly when asked to make difficult decisions, and would often end up relying on gut feeling, leaving even close allies in the dark about his plans. His "unreliability had those who worked with him pulling out their hair," as his confidant Ernst Hanfstaengl later wrote in his memoir *Zwischen Weißem und Braunem Haus*. This meant that rather than carrying out the duties of state, they spent most of their time in-fighting and backstabbing each other in an attempt to either win his approval or avoid his attention altogether, depending on what mood he was in that day.

There's a bit of an argument among historians about whether this was a deliberate ploy on Hitler's part to get his own way, or whether he was just really, really bad at being in charge of stuff. Dietrich himself came down on the side of it being a cunning tactic to sow division and chaos—and it's undeniable that he was very effective at that. But when you look at Hitler's personal habits, it's hard to shake the feeling that it was just a natural result of putting a work-shy narcissist in charge of a country.

Hitler was incredibly lazy. According to his aide Fritz Wiedemann, even when he was in Berlin he wouldn't get out of bed until after 11:00 a.m., and wouldn't do much before lunch other than read what the newspapers had to say about him, the press cuttings being dutifully delivered to him by Dietrich. But he didn't even enjoy being in Berlin, where people kept on trying to get him to do stuff: he'd take any opportunity to leave the seat of government and go to his private country retreat in the Obersalzberg, where he'd do even less. There, he wouldn't even leave his room until 2:00 p.m., and he spent most of his time taking walks, or watching movies until the small hours of the morning.

He was obsessed with the media and celebrity, and often seems to have viewed himself through that lens. He once described himself as "the greatest actor in Europe," and wrote to a friend, "I believe my life is the greatest novel in world history." In many of his personal habits he came across as strange or even childish—he would have regular naps during the day, he would bite his fingernails at the dinner table and he had a remarkably sweet tooth that led him to eat "prodigious amounts of cake" and "put so many lumps of sugar in his cup that there was hardly any room for the tea."

He was deeply insecure about his own lack of knowledge, preferring to either ignore information that contradicted his preconceptions, or to lash out at the expertise of others—he was said to "rage like a tiger" if anybody corrected him. "How can one tell someone the truth who immediately gets angry when the facts do not suit him?" lamented Wiedemann. He hated being laughed at, but enjoyed it when other people were the butt of the joke (he would perform mocking impressions of people he disliked). But he also craved the approval of those he disdained, and his mood would quickly improve if a newspaper wrote something complimentary about him.

Little of this was especially secret or unknown at the time. It's why so many people failed to take Hitler seriously until it was too late, dismissing him as merely a "half-mad rascal" or a "man with a beery vocal organ." In a sense, they weren't wrong. In another, much more important sense, they were as wrong as it's possible to get. Hitler's personal failings didn't stop him having an uncanny instinct for political rhetoric that would gain mass appeal, and it turns out you don't actually need to have a particularly competent or functional government to do terrible things.

We tend to assume that when something awful happens there must have been some great controlling intelligence behind it. It's understandable: How could things have gone so wrong, we think, if there wasn't an evil genius pulling the strings? The downside of this is that we tend to assume that if we can't immediately spot an evil genius, then we can all chill out a bit because *everything will be fine.*

But history suggests that's a mistake, and it's one that we make over and over again. Many of the worst man-made events that ever occurred were not the product of evil geniuses. Instead, they were the product of a parade of idiots and lunatics, incoherently flailing their way through events, helped along the way by overconfident people who thought they could control them.

6 GOVERNMENT POLICIES THAT DID NOT WORK OUT WELL

Poll Tax

The smartest minds in Margaret Thatcher's government came up with what they thought a fairer tax: one where everybody, rich or poor, paid the same. It led to widespread nonpayment, large-scale riots and eventually Thatcher was forced to resign.

Prohibition

America's efforts to ban the drinking of alcohol between 1920 and 1933 did lead to fewer people drinking—but it also allowed organized crime to monopolize the alcohol industry, making crime soar in many places.

The Cobra Effect

As pest control in Delhi, the British government offered a bounty for dead cobras. So people simply bred cobras to claim the bounty. So the British dropped the bounty. So people turned the worthless cobras loose. Result: more cobras.

The Smoot-Hawley Tariff Act

As the Great Depression started to bite in 1930, the US introduced large tariffs on imports to try and prop up domestic industries. Instead, the resulting trade war only worsened the global depression.

The Duplessis Orphans

In Quebec in the 1940s and 1950s, the government offered church groups subsidies to care for both orphans and the mentally ill. But the psychiatric payments were double that for orphans—so thousands of orphans were falsely diagnosed as mentally ill.

Hoy No Circula

In 1989, Mexico City tried to reduce air pollution by banning particular cars from being driven on certain days. Unfortunately, rather than taking the bus, people just bought more cars so they'd always have one that was legal to drive.

6

WAR. HUH. WHAT IS IT GOOD FOR?

Humans are very keen on war. It is, in many ways, our "thing." The oldest evidence in the archaeological record for organized mass violence dates back to around 14,000 years ago, at Jebel Sahaba in the Nile Valley, although let's be honest, we've probably been having fights of some kind a lot longer. Meanwhile (as mentioned a couple of chapters ago), evidence from Oaxaca in Mexico suggests that pretty much as soon as people started living in villages, one village would try to raid another, and things would quickly escalate from there.

It's estimated that between 90 and 95 percent of all known societies have engaged in war on a fairly regular basis; the few that mostly manage to avoid it tend to be relatively isolated ones that stuck with a nomadic, foraging or hunter-gatherer lifestyle.

There is one notable historical exception to this, though—the Harappan civilization that existed in the Indus Valley from 5,000 years ago, stretching across parts of modern-day Afghanistan, Pakistan and India. Rising at around the same time as those in Mesopotamia and Egypt, the Harappan civilization was an advanced society with a population in the millions. It had major cities that show sophisticated urban planning and boast things like plumbing and toilets and public baths, and it was home to a culture that produced innovative technology and art that was traded far and wide. And, it seems, it had basically no war. At all. Archaeologists have been excavating the remains of Harappan cities for a century now and they've found little evidence of settlements being raided or destroyed, only a few examples of significant fortifications or defenses, no depictions of warfare in Harappan art and nothing that suggests the existence of an army or large collections of military weapons. (And interestingly, unlike other comparable civilizations of the same period, they also haven't found much in the way of monuments to great leaders.)

This occasionally leads to a portrayal of the Harappans as some sort of idealized proto-hippies, which is a nice idea but probably closer to wishful thinking than reality. While they do seem to have been a pretty chill society who got on well with their neighbors, they also had the advantage of being geographically well protected from anybody who might have wanted to invade them, which certainly makes it a lot easier to not have wars. And it is of course possible that we simply haven't found the evidence of war *yet*; if so, it wouldn't be the first time that a civilization gained a reputation for pacifism only for later discoveries to completely ruin that reputation. The Harappan writing system still hasn't been properly deciphered, so maybe one day we'll decode it and find it says, "LOL, let's hide all our war stuff to really confuse the archaeologists."

Still, for now it does seem that, at the exact same time other early civilizations were really leaning into the whole war and conquest thing, Harappan society managed to exist at its peak for 700 years without being troubled in any major way by external conflict. And then, for uncertain reasons, the Harappan civilization just kind of...fades out of history. Its people start moving away from the cities and returning to the countryside. The change in climate around 2200 BCE, which caused the decline of several other early civilizations, would have made the valley increasingly arid and less fertile; overpopulation and overfarming may have led to food shortages; and like all dense urban populations, they were more vulnerable to infectious diseases. Whatever the cause, by 3,500 years ago the cities had been almost entirely abandoned, and this brief no-war blip in humanity's history was over. Meanwhile the rest of the world's civilizations continued to grow and carry on doing war stuff.

(The unsettling possibility here is that the Harappans' key mistake was not having wars, and that civilization actually needs war to sustain itself. There's your cheerful thought for the day.)

Right now, we're lucky enough to be living in a relatively peaceful period of history, though even so, you might have noticed we haven't exactly been short of a war or two. The annual death tolls from wars around the world have been on a downward slope for several decades now, which some writers have suggested shows that we have in fact entered a new era of peace and rationality and international friendship. In honesty, though, it's probably a bit soon to be claiming that: after all, the slope is heading down from history's biggest peak in World War II. Humanity could just be having a bit of a breather before starting up again.

In a book about failure, I hope it goes without saying that all wars are, to some extent, enormous failures on *somebody's* part. But in addition to them being very bad things in their own right, the chaos and tunnel vision and general macho nonsense of wars also really heighten humanity's innate capacity for failing hard in many other ways. War is a collective rush of blood to the head; to put it another way, it's fuck-up central.

Nowhere is this clearer than in the justly celebrated Battle of Cádiz, which should possibly be more accurately renamed the Piss-Up of Cádiz. In 1625, the English decided that they wanted to fuck up the Spanish good and proper. King James VI and I (of kingdom-unifying, Bible-commissioning and witch-hunting fame) had just died, leaving his large adult son Charles I in charge. Charles—demonstrating all the tact and judgment that would eventually see him minus a head—had held a grudge against Spain ever since they wouldn't let him marry one of their princesses, and he wanted some payback. So he and his pals decided to kick it old-school and mount some piratical raids to steal all the gold and silver the Spaniards were shipping back from the Americas.

In November that year, 100 ships and 15,000 troops of a joint English and Dutch expeditionary force sailed into the Bay of Cádiz in southwestern Spain. They were there to plunder, and they weren't taking no for an answer. Granted, they were only in Cádiz because they'd been so disorganized and delayed that they'd completely missed the Spanish fleet and its treasure on its return from America, but still. It was payback time.

Unfortunately, even before they reached Cádiz it had become apparent they'd not brought enough food or drink with them. So when the invading forces landed, the expedition's commander, Sir Edward Cecil, decided to let his starving troops prioritize finding sustenance over, you know, fighting any battles. Naturally, his troops immediately did what English people abroad always do: they made a beeline for Cádiz's wine stocks. And proceeded to get well and truly wankered.

Upon realizing that his entire army was twatted, Cecil took the reasonable decision to abandon the plan entirely, and ordered his men to retreat to the ships and slink home in shame.

Most of them did, eventually, but about 1,000 were so drunk that they just stayed lounging around Cádiz until the Spanish forces turned up and executed them all.

And that's how England failed to invade Cádiz.

The English exploits in Cádiz often appear on lists of history's greatest military failures—but to be entirely honest, if you ignore the bit about people being executed, it actually sounds pretty great. Turn up, don't eat enough, get riotously pissed and lose a couple of your mates along the way: that's a classic holiday. If instead of having wars, we just sent large groups of people to each other's countries to drink loads of their wine and aimlessly wander around their towns on a regular basis, then the world would probably be a much, much happier place. Although now I've written that, it occurs to me that's essentially what the EU is.

Alcohol, you'll be astonished to learn, plays a leading role in a number of the dumbest moments to have graced the battlefield—as is the case with the Not-Really-a-Battle of Karansebes in 1788. This was impressive for the way the Austrian army managed to suffer heavy losses in battle despite the fact that their opponents never even turned up. In fact, their enemy (they were fighting the Ottoman Empire at the time) didn't actually know the battle had happened until they came across the aftermath a little while later.

Exactly *what* happened is, uh, somewhat murky. What's fairly clear is that the Austrian army was retreating at night through the town of Karansebes (in modern-day Romania), keeping a wary eye out for the pursuing Turks. At this point, accounts of the incident diverge. In one telling, a unit of local troops from the Romanian region of Wallachia started to spread rumors that the Turks had arrived, in order to cause confusion so they could loot the baggage train. In another telling, a group of cavalry officers met a Wallachian farmer with a cartload of brandy and decided they'd had a long, hard day's riding and deserved some downtime. After a while, a group of infantry turned up and pointedly inquired whether the cavalry were planning on sharing the brandy with their foot-soldier brethren, at which point things got…rowdy.

Whatever the cause (the divergent tales have the definite ring of every unit of the army trying to blame another unit of the army), most sources seem to agree that things come to a head when somebody fires a shot into the air, and then somebody else starts shouting, "The Turks, the Turks!" The (quite possibly drunk) cavalrymen think it's serious, and so naturally they start riding around shouting, "The Turks, the Turks!" as well. At which point everybody panics like fuck and starts trying to flee from the imaginary Turkish forces. In the darkness and confusion and probably drunkenness, two columns of troops cross each other, both mistake the other for the dreaded enemy and they begin firing wildly on each other.

By the time everybody has worked out that there aren't actually any Turks attacking them, quite a lot of the Austrian army has run away, wagons and cannons have been overturned and the bulk of their supplies have been lost or ruined. When the Turkish army do turn up the next day, they discover a number of dead Austrians and the scattered remains of their camp.

Estimates of the losses vary quite dramatically. One source simply says "many" were dead and wounded; another says 1,200 were injured; while the Austrian leader Emperor Joseph II underwhelmingly claims in a letter that they lost "not only all the pots and tents…but also three pieces of artillery." The most famous accounts of the battle put the death toll as high as 10,000, but that's almost certainly a number some bloke invented to make the story sound better. In conclusion: something happened, some people may or may not have died, but everybody agrees that it was extremely stupid.

I think this is what's referred to as "the fog of war."

Another fine example of effectively managing to defeat yourself came during the Siege of Petersburg in the American Civil War, when the Union troops turned a tactical triumph into a humiliating setback in a particularly inventive way. They had Confederate forces pinned down in the fort, and spent a month preparing for the coup de grâce that would breach their walls in one dramatic maneuver, by digging a 500-foot mine shaft directly underneath the Confederate fort and planting an awful lot of explosives there.

When they blew up the wall in the early hours of the morning of July 30, 1864, the size of the explosion seems to have taken everybody by surprise. It killed hundreds of Confederate troops and left an enormous crater, 170 feet long and 30 feet deep. After about ten dazed minutes spent staring at it in shock, the Union forces attacked—although, unfortunately, they weren't the troops who'd trained for days in the tactics they'd use to storm the fort once the wall was breached. That's because the soldiers who'd been trained were black, and at the last minute the commander of the Union army instructed his underlings to swap them out for white soldiers because he was worried how it would look. And so the white troops rushed toward the Confederate position—and ran straight into the crater.

It's possible they thought the crater would provide good cover. It didn't. Once the Confederate soldiers had reorganized after the shock of the explosion, they found themselves surrounding a very large hole full of opponents who couldn't get out. Union reinforcements kept on arriving, and for some reason decided to join their comrades in the crater. The Confederate commander later described it as a "turkey shoot."

The key lesson in military tactics we can learn from this is: don't walk into big holes in the ground.

Another essential lesson for any budding military strategists is that communication in wartime is vitally important. That's something that the Pacific island of Guam learned during the Spanish–American war of 1898, when their colonial masters in Spain forgot to tell them that there was a war happening at all.

As a result of this oversight, when a small fleet of US warships steamed up to a suspiciously underdefended Guam and fired 13 shots at the old Spanish fort of Santa Cruz, Guam's dignitaries reacted by rowing out to the battleships, thanking the Americans for the generous greeting salute and apologizing that it would take them a while to return the courtesy because they'd need to move their cannons over from another part of the island.

After a few awkward moments, the Americans explained that they hadn't been saying hello, they'd actually been trying to have a battle, because there was a war on. The dignitaries, who were somewhat miffed to find they were now prisoners of war, explained they hadn't received any messages from Spain in over two months and were completely in the dark about the whole war thing. They went off to have a bit of a debate about what to do, while one of the local merchants stuck around for a chat because it turned out he was an old friend of the American captain.

Guam officially surrendered a few days later, and has been an American territory ever since.

As a species, we're not great at the "don't repeat the mistakes of history" thing. But few examples are quite as glaring as the fact that in 1941, Hitler very precisely copied Napoleon's fatal mistake from 129 years earlier, one which in both cases utterly screwed their previously quite successful plans to conquer all of Europe. That mistake, of course, was trying to invade Russia.

History's only truly successful large-scale invasion of Russia—or rather Kievan Rus', as Russia didn't exist then—was by the Mongols, and they were fairly unique as these things go (as we'll see in a few chapters). The Poles managed it for a short while (and even held Moscow for a couple of years) but were still ultimately driven back, while it went extremely badly for Sweden the one time they tried it, ending in a defeat that helped to effectively finish off the Swedish Empire. Basically, "don't do it" is the lesson we're learning here.

Of the two leaders, Napoleon's rationale for going ahead with the plan was marginally better than Hitler's. For starters, he didn't have the example of Napoleon's previous failure as a useful guide. He also had every reason to be confident of victory, given the Harlem Globetrotters–level winning streak his Grande Armée were on up to that point. Additionally, he had some legitimate mild beef with Tsar Alexander, who he thought was undermining his economic blockade of the British, the only other major holdout to his total European conquest. Granted, quibbling over a trade embargo isn't really a great reason to begin hostilities with a massive country. If Napoleon made one key mistake, it's that his methods of getting his way pretty much started and ended with "have a war." Diplomacy and negotiation were not really his strong suit.

With his decision that he was going to invade *somebody* effectively premade, Napoleon must have thought Russia seemed like a safer bet than Britain, because at least it was overland. And knowing that the Russian climate effectively only gave him three months of invading time, he came up with a strategy: head straight for Moscow and force the Russians into a pitched battle, which he would win thanks to having an army who were actually motivated and good at their jobs rather than a load of mercenaries being ordered around by aristocrats.

Unfortunately, this was one of those plans that sounds great when you say it, but entirely relies on your opponents doing exactly what you want them to do. Instead, the Russians pretty much allowed them to march in. They retreated and retreated, avoiding major battles wherever possible, all the while scorching the earth to deny the French supplies, and simply waited for the winter to arrive and do the job for them. By the time Napoleon realized what the game was, it was too late to get out before the cold bit, and the French were forced into the long, grim death march home with a shattered army. The rest of Europe suddenly saw weakness where previously there had only been strength, and that was the beginning of the end for Napoleon.

In 1941, Hitler was in a similar position: having also discovered the difficulties of invading Britain, due to the whole island business, he, too, decided he had a narrow summer window in which to invade the Soviet Union as an alternative. Granted, he actually had a nonaggression pact with the Soviets at the time, but on the other hand he was a Nazi and they were communists and so he hated them.

Hitler actually studied Napoleon's strategy and thought he'd come up with a clever plan to avoid the same mistakes. Rather than sending all his forces directly at Moscow, he'd divide them into three, attacking Leningrad and Kiev as well as the Soviet capital. And unlike Napoleon, he wouldn't retreat at the first sign of winter, but stand and fight. These were both disastrous choices. What he didn't spot was that although the tactics might be different, the basic plan (strike quickly and decisively, win big battles easily, assume this all leads to the swift collapse of your opponents) remained the same. As did its flaws (relies on opponents to follow your script, no plan B when they mysteriously don't, still completely ignoring the thing about the Russian winter).

There were plenty of people in the German High Command who could have pointed these flaws out to Hitler, but as soon as he caught a whiff of dissent or skepticism, he would keep them in the dark about his plans, or flat-out lie to them. It was a decision-making process based in equal parts on hubris, wishful thinking and sticking his head in the sand.

The strategy's flaws were the same as Napoleon's, and the outcome was roughly the same, too, albeit even more deadly this time around. The Germans made major territorial gains and won some battles, but the Soviets didn't collapse like the script demanded. They used scorched-earth tactics and kept the Germans bogged down until winter, at which point it turned out that they didn't have the right clothes, enough supplies or indeed

German retreat in Russia, 1944

antifreeze for their tanks. Hitler's orders to stay and fight in the bitter cold rather than retreating didn't bring him any greater success; it just killed more of his soldiers. For the second time, an army that had conquered much of continental Europe was catastrophically weakened by a needless invasion of Russia, and the tide of the war was turned.

As a bonus, at about the same time, Germany's allies in Japan were busy launching their own badly thought-through attack on Pearl Harbor that needlessly dragged a superpower into a war they'd been trying to stay out of. Without those two woefully poor choices, the Axis powers might have won. Proving that just sometimes, humans' extremely poor decision-making skills can work out for the best in the long run (at least assuming you're not a fan of Hitler).

With the Americans and the Japanese now engaged in battle on the Pacific, there was a chance to prove that the fog of war can involve very literal fog as well as the metaphorical kind. That was the case on Kiska, a barren but strategically important island in the North Pacific that lies about halfway between

Japan and Alaska (of which it is a very remote part). It was one of two islands captured by the Japanese in 1942 at the height of World War II, which freaked the Americans out because it was the first time since they fought the British in 1812 that their territory had been occupied. Even if the territory was extremely small and far away.

In the summer of 1943, 34,000 US and Canadian troops prepared to try and recapture Kiska. They were still bruised and wary from the experience of retaking nearby Attu Island, a brutal and bloody affair in which the Japanese forces had fought to the death. The operation's commanders were certain that the battle for Kiska would be every bit as ferocious. When they landed on August 15, the Allied forces found Kiska shrouded in thick, freezing fog. In hellish conditions of bitter cold, wind and rain with zero visibility, they blindly picked their way step by step across the rocky terrain, trying to avoid mines and booby traps, while all the time bursts of gunfire from unseen enemies lit up the fog around them. For 24 hours they dodged sniper fire and painfully inched their way up the slope toward the center of the island, accompanied by muffled explosions from artillery shells, the staccato sound of nearby firefights and indistinct shouts trying to convey orders or rumors of Japanese forces close by.

It was only the next day, as they counted their losses—28 dead, 50 wounded—that they realized the truth: they were the only ones there.

The Japanese had actually abandoned the island almost three weeks earlier. The US and Canadian forces had been shooting at each other.

This would probably go down as an unfortunate but understandable mistake, except for one thing. Their aerial surveillance team had actually told the operation's leaders weeks before the landing that they'd stopped seeing any Japanese activity on the island, and thought it had probably been evacuated. But after the experience on Attu, the leaders had convinced themselves that the Japanese would never retreat, and so dismissed the surveillance reports. It was confirmation bias run wild. They were so certain, they even turned down the offer to fly a few more surveillance missions just to double-check. There's probably a lesson there about not making assumptions.

Two years later, in April 1945—mere weeks before the end of the war—the German U-boat *U-1206* was nine days into its maiden active voyage, patrolling the waters off the northeast coast of Scotland. It was a state-of-the-art vessel, fast and sleek and high-tech, and, crucially, with a fancy new type of toilet that would shoot human waste out into the sea rather than storing it in a septic tank.

The only downside of the toilet was that it was unexpectedly complicated to use. So much so that on April 14, the captain was forced to call an engineer because he couldn't work out how to get the thing to flush, which is probably not the sort of thing you want when you're trying to maintain an air of authority. Unfortunately, the engineer wasn't any better at toilet-flushing. In trying to operate the mechanism, he somehow turned the wrong valve—which quickly caused the cabin to start flooding with a deeply unpleasant mixture of seawater and human excrement.

No, I don't know who decided "let's put a valve in the toilets that looks *an awful lot* like the flush mechanism but instead lets seawater pour into our big Nazi submarine," but presumably they were from the same school of thought as the guy who put that exhaust port in the *Death Star*.

The flooding of the cabin with a pungent cocktail of feces and brine would have been bad enough, but things got significantly worse when the sewage leaked down a deck onto the submarine's batteries, which the boat's designers had helpfully installed directly below the toilet. This caused the batteries to start spewing out large amounts of deadly chlorine gas, presenting Captain Schlitt with no option but to surface—where he was promptly attacked by the RAF, forcing him to abandon ship entirely and scuttle the vessel. Leaving *U-1206* with the unfortunate legacy of being the only craft in World War II to have been sunk by a poorly thought-out toilet.

There are valuable lessons here about the paramount importance of user-interface design in high-pressure environments, and the necessity of physically separating pieces of mission-critical infrastructure, but to be perfectly honest, I only included it because it's really funny.

Having a plan is obviously crucial to military success. But sometimes it's possible for a plan to be too cunning and devious for its own good. If you've ever played chess against somebody much, much better at it than you, you're probably familiar with how it goes: you spend ages trying to maneuver them into an extremely clever trap, only to realize that they've anticipated every move and you've actually defeated yourself. That's basically what French general Henri Navarre did in Vietnam, except he did it with people rather than chess pieces. Like his earlier compatriot Napoleon, he hatched a plan that was perfect just so long as his opponents did exactly what he wanted them to.

It was 1953, and Navarre's goal was to inflict a crushing and humiliating defeat on the communist Viet Minh forces (who were doing an annoyingly good job of rebelling against colonial rule in French Indochina) in order to weaken their hand in the imminent peace negotiations. So he decided to set an extremely clever trap for them. He built a major new French base in a remote area, threatening Viet Minh supply lines, and tried to draw them into a fight. The base at Điện Biên Phủ was surrounded by mountains covered in thick jungle, which gifted the Vietnamese the advantage of cover and high ground. The French were a long way from reinforcements. It was simply too tempting a target for the Viet Minh to resist. But (the plan went) superior French technology would defeat them easily: France's air dominance would allow them to fly in supplies, while French firepower would triumph in the battle, as transporting heavy artillery through the jungle would be impossible for the Viet Minh. Excellent plan. Navarre had his men set up the base, and then waited.

And waited. For months, nothing happened. No attack came. What were the Viet Minh doing?

Turns out that what they were doing was transporting heavy artillery through the jungle. A combination of Vietnamese troops and local civilians spent those months disassembling their weapons, carrying them piece by backbreaking piece across miles of thickly forested mountain to Điện Biên Phủ and then putting them back together. After that, they simply waited for the rainy season to start, and once the French forces were stuck in the mud and the French planes couldn't see where to drop supplies, they attacked. Navarre's men, who had been expecting doomed suicidal foot charges by peasants carrying outdated rifles, were surprised to come under sustained bombardment from advanced artillery that wasn't supposed to exist.

The French troops held out under siege for two months before they were overrun. The scale and the manner of the defeat was so crushing and embarrassing that the French government fell, and the Viet Minh helped secure independence for what became known as North Vietnam. After that, the rest is a familiar story: with Vietnam divided into two states, the remnants of the Viet Minh that remained in South Vietnam turned into the Viet Cong, who quickly began a violent insurgency against the southern government. The US decided to get involved to support their allies in the south, because of the whole Cold War anticommunism thing, whereupon Uncle Sam turned out not to be much better than the French at fighting basically the same war. The ensuing Vietnam War lasted for almost two decades, and somewhere between 1.5 million and 3 million people died. All of which happened, in part, because Henri Navarre came up with an extremely clever trap.

But in the annals of military failures, it's a different front in the attempt to heat up the Cold War that provides the most indelible example—one in which the cognitive biases of a small group of people saw a superpower humiliated by a minnow.

Bay of Pig's Ear

The American debacle when they tried to invade Cuba via the Bay of Pigs isn't just a classic example of groupthink in action—it's literally where we get the word from. It was coined by psychologist Irving Janis based in large part on his study of how the Kennedy administration managed to get things so wrong.

The Bay of Pigs operation was almost certainly the most humiliating incident in America's long-running and hilarious string of failures to overthrow the government of a small island situated right on its doorstep, although in fairness, it might not be the weirdest. (That would probably be the CIA buying up a large number of mollusks in an attempt to assassinate a scuba-diving Fidel Castro with a booby-trapped shellfish.)

The basic plan went like this: the US would train up a group of anti-Castro Cuban exiles who would mount an invasion with American air support. Upon seeing their initial easy victories against the ramshackle Cuban military, the people of the island would greet them as liberators and rise up against the communists. Simple. It was what they'd already done to Guatemala, after all.

The wheels started to come off when John F. Kennedy beat Richard Nixon to the presidency. The plan had been developed with the assumption that Nixon, previously the vice president and a supporter of the scheme, would be the new man in the Oval Office. Kennedy was considerably less gung ho and, not unreasonably, worried about starting a war with the Soviets, so he insisted on some changes: US backing for the operation had to remain completely secret (so no air support), and the landing site had to be changed to somewhere far from large civilian populations, somewhat undermining the "trigger a popular uprising" element.

At this point it should have been clear that the already fairly optimistic operation should just be scrapped, because it didn't make even a lick of sense anymore. And yet everybody just carried on as though it did. Questions weren't asked, assumptions went unchallenged. The historian Arthur Schlesinger, an adviser to the Kennedy administration who opposed the plans, later said that the meetings about it took place in "a curious atmosphere of assumed consensus," and that even though he thought the plan was stupid, in the meetings he found himself staying quiet. "I can only explain my failure to do more than raise a few timid questions by reporting that one's impulse to blow the whistle on this nonsense was simply undone by the circumstances of the discussion," he wrote. In fairness, we've all been in meetings like that.

When the attack actually started, in April 1961, pretty much everything that could go wrong did go wrong. Without the US air force to take out Castro's air force, the job was left to Cuban exiles flying bombers out of Nicaragua disguised to look like Cuban planes. The plan was to have one plane very publicly land in Miami, and the pilot to announce to the world that he was a defector from the Cuban military who had decided to bomb the air bases himself. This cunning ruse lasted about as long as it took people to notice that the plane wasn't actually the same type used by the Cubans.

The landing party, which was supposed to arrive secretly under the cover of darkness, was quickly spotted by some local fishermen, who, instead of greeting them as liberators, raised the alarm and then went to shoot at them with rifles. ("We thought, this is the invasion, boys, be careful! They are trying to invade," one of the fishermen, Gregorio Moreira, recalled to the BBC on the fiftieth anniversary of the invasion.) The invaders quickly discovered that the beach they were supposed to take over the country from was actually quite hard to get off, and only got harder when large parts of the Cuban army (who turned out to be quite efficient, not ramshackle at all) quickly showed up and started shooting at them. Oh, as did a plane from the Cuban air force, which it appeared hadn't been destroyed by the unconvincing fake bombers, after all.

At this point, the beach force could really have done with some air support, but by now Kennedy was so unnerved by the fact everybody had seen through the "defecting Cuban pilots" trick that he refused to authorize it. So they remained stuck on the beach for several days, fighting an increasingly desperate defense with ammunition running low.

Three days into the abortive invasion, it was clear that they were never getting off the beach without a dramatic intervention, and so finally Kennedy did an about-face and gave authorization for air support. But at this point, the Cuban pilots felt so betrayed by the way the mission had gone that they refused to fly. So the US dumped all pretense of not being involved, and drafted in members of the Alabama National Guard to fly the disguised bombers, which would be supported by a bunch of regular, extremely nondisguised US fighter planes. This might have given the landing party on the beach a fighting chance, except that in the final piece of glorious incompetency, they forgot that there's a time difference between Nicaragua, where the bombers were, and Miami, where the fighters were, so the two sets of planes didn't even manage to meet up. Several of them were shot down.

The whole thing ended with the US a global laughingstock, Fidel Castro more firmly entrenched in power than ever and over 1,000 invading troops captured, who a few years later the US would have to pay a ransom of over $50 million to free.

On the plus side, Kennedy learned from the decision-making failures—which might possibly have saved everybody in the world when it allowed cooler heads to prevail during the Cuban

Missile Crisis the following year. And thankfully, such was the impact, the United States never again let itself get into a situation where its leaders would allow groupthink to push them into a poorly thought-out invasion based on shoddy intelligence with no clear plan or exit strategy.

Oh.

6 OF THE MOST POINTLESS WARS IN HISTORY

The War of the Bucket

An estimated 2,000 people died in this 1325 war between Italian city-states Modena and Bologna, which began when some Modenese soldiers stole a bucket from a well in Bologna. Modena won, and promptly stole another bucket.

The Anglo-Zanzibar War

The shortest war in history, at under three quarters of an hour. A Zanzibari sultan the British didn't approve of claimed the throne, then barricaded himself in the palace, which the British proceeded to shoot at for a total of 38 minutes before he fled.

The Football War

In 1969, the long-simmering tensions between El Salvador and Honduras spilled over into actual war—largely sparked by violence during a series of tense World Cup qualifiers between the two countries. (El Salvador won the football; war was a draw.)

The War of Jenkins's Ear

A war between Britain and Spain that lasted for over a decade and cost tens of thousands of lives started because in

1731 some Spanish privateers cut off a naval captain's ear. By the time it was over it had expanded into the War of Austrian Succession, which involved just about every major country in Europe.

The Chamber Pot Rebellion

Robert Curthose was the eldest son of William the Conqueror who was moved to open rebellion against his father when William didn't punish his two younger sons sufficiently after they tipped a full chamber pot over Robert's head.

The War of the Golden Stool

A war between the British Empire and the Ashanti people of West Africa that started after the British governor threw a strop about the "ordinary chair" he was given and demanded to sit on the Golden Stool—a sacred throne nobody was allowed to sit on. The British won the war, but he never got to sit on the stool.

SUPER HAPPY FUN COLONIALISM PARTY

The human compulsion to explore, to always seek out, new horizons is one of our defining characteristics. It's why our species and its close cousins spread around the world multiple times in an evolutionary blink of the eye. And it's the driving force that gave shape to the modern world—which is the dumb, chaotic and frequently wildly unfair product of millennia of migration and trade, colonization and war.

It was that urge to explore that drove Christopher Columbus to set sail into the vast, empty blue waters of the Atlantic in 1492, only to end up crashing onto a load of rocks a few months later like an idiot.

That year was right near the beginning of what is commonly called the "Age of Discovery," although it was only really discovery if you weren't one of the people already living in the places being discovered. The overland trade routes between Europe and Asia that had been nice and easy to travel when the Mongol Empire stretched across much of Eurasia (again, more on that in a bit) were now blocked, thanks to a combination of the Black Death and the rise of the Ottoman Empire. And so Europe, buzzing with new technology and knowledge and hungry for riches, set its eyes to the sea instead. And what began as a drive for trade in Asia, Africa and the newly discovered Americas would quickly turn into missions of occupation and conquest.

Pretty much everybody knows that Columbus discovered (well, "discovered") the Americas by accident, running into the Caribbean by mistake while he was looking for a shortcut to India that didn't involve going around the southern cape of Africa. But there are also a lot of misconceptions around exactly *what* his mistake was.

In one common telling, he was essentially proved right, because he had confidence in his heretical theory that the world was round; meanwhile the credulous fools back home believed he was doomed to sail off the edge of the world. This is, I'm sorry to say, utter balls. In fact, pretty much every educated person in Europe at the time (and most of the uneducated ones, too) was fully aware that the world was a globe, and they'd known that for a very long time. It was such common knowledge that over 200 years before Columbus's journey, the theologian Thomas Aquinas was casually using it in his writing as an example of something that everybody accepted to be true. Given that to this day there's a persistent minority of people who still doubt the official story put out by Big Globe, the flat-earth theory may well be as popular right now as it was in the fifteenth century. In 2019, a group of flat-earthers are planning to organize a cruise for the flat-earth community, which should be an exciting opportunity for them to test out their theories. Well done, everybody, for that.

So no, it wasn't disagreement about the roundness of the earth. The skepticism about Columbus's venture came from an entirely different source. The problem was that Christopher Columbus had utterly messed up his units of measurement, so got his sums completely wrong.

His entire plan for the mission was based on his personal calculations of two things: how big the earth was, and how big Asia was. On both of these, he was wildly off. For one, he decided that Asia was an awful lot longer than it actually is (and it is pretty long), and that, as such, with a fair wind he'd eventually find Japan several thousand miles to the east of Japan's actual location. But even worse, he based his calculations of the globe's circumference on the work of the ninth-century Persian astronomer Ahmad ibn Muhammad ibn Kathir al-Farghani. This wasn't a great start, as there had been more accurate estimates around since the Greek mathematician Eratosthenes of Cyrene pretty much nailed it 1,700 years earlier. But even that wasn't Columbus's biggest mistake.

His biggest mistake was his assumption that when al-Farghani said "mile," he was obviously talking about the Roman mile, which was around 4,850 feet. That is not what al-Farghani was talking about. He was talking about the Arabic mile, which is somewhere between 6,500 and 7,000 feet. So when al-Farghani said that something was a certain number of miles away, he actually meant a much, much bigger distance than Columbus realized.

Fans of the movie *This Is Spinal Tap* will be familiar with what Christopher did there. He confused one unit of measurement with a completely different unit of measurement, and so came up with a model that was ridiculously small. Columbus thought that the world was only about three quarters of the size it actually is. Combined with his decision to relocate Japan by several thousand miles, the upshot was that he thought he needed to pack supplies for a much shorter journey than the one he faced. Plenty of his contemporaries were all like, "Think you've got the world the wrong size there, Chris," but he remained convinced by his calculations. So on the whole, it was pretty lucky for him that he bumped into the Caribbean when he did. (Nobody had really given serious thought to the possibility that there was a whole extra continent in exactly the place that Asia wasn't.)

It's probably worth adding that his erroneous assumption about which kind of mile al-Farghani was talking about reflects some pretty Eurocentric thinking on Columbus's part! But let's be honest: this was nowhere near the worst thing Christopher Columbus did because his thinking was too Eurocentric.

It's tempting to wonder how different world history might have been if Columbus had been better at math and so never set out on his voyage. The answer is: probably not much, except maybe some more people might speak Portuguese now. The Portuguese were the best sailors and navigators in Europe at the time (Columbus's expedition was only funded by Spain because Portugal rejected it first, on the grounds that they knew perfectly well he'd fucked the math), and they would land on various bits of the Americas in the following years. Pedro Álvares Cabral reached Brazil in 1500. One year later, the Corte-Real brothers reached either Labrador or Newfoundland, where in a sign of things to come they promptly kidnapped 57 natives and sold them as slaves.

In fact, the thing that would really have made a difference to the history of Old World–New World relations is if anybody, literally anybody, had been able to constrain their natural impulse to murder or kidnap the first people they met. A whole five centuries before Columbus, the Vikings were the actual first Europeans to establish a settlement in the Americas, with Leif Eriksson setting out from the Viking colony on Greenland and coming upon what they decided to call Vinland ("Wine

Land," which was likely modern-day Newfoundland). Compared to barren and extremely un-fun Greenland, the woods and fruits of Vinland should have been great news for the Vikings, and they did indeed establish a trading colony there for some years. Unfortunately, their prospects for trade with the locals of Vinland (likely the Thule people, or Skraelings as the Vikings called them) were somewhat diminished by what happened the first time they encountered each other.

This was the first meeting between Europeans and Americans in recorded history, and it went something like this: the Vikings found a group of 10 natives sleeping under their upturned canoes, and so they murdered them.

For fuck's sake, guys.

Unsurprisingly, after that the locals weren't terribly keen on trading with the Vikings, and skirmishes between the two groups were common—including one battle in which the fearsome Vikings, armed with swords, were almost defeated by "a pole with a huge knob on the end" (likely an inflated animal bladder), "which flew over the heads of the men and made a frightening noise when it fell." The Vikings were so scared of the novelty balloon that they would have lost the fight if Freydis Eriksdottir, Leif's sister, hadn't alarmed the Skraelings in turn by exposing her breasts.

As a result of this and other less weird battles, the Vinland settlement never really got off the ground. The Vikings of Greenland abandoned it after a decade or two. What's more, the Greenland settlement itself—which only came into existence in the first place because Erik the Red had been exiled there for murdering people—gradually withered and died over the following centuries, as the rest of the Vikings back in Nordic lands largely stopped paying attention to it.

If things had gone a bit differently on Vinland, ideally with less murder, then history might genuinely have taken a different course. An established trading route between the Americas and Europe, with all the exchange of knowledge and skills that can lead to, might have resulted in a more gradual exposure between the two populations. It could have meant that the gap in technology and military might that made European colonization in the sixteenth century such a one-sided affair would have been less dramatic. (It might also have given the Americans more time to slowly develop resistance to Old World infectious diseases, rather than getting hit with a mother lode of them all at once.)

Equally, things might have been different if Abubakari II, the ruler of the Malian empire in the fourteenth century, had come back from his voyages. The emperor of one of the world's largest and wealthiest empires at the time, spanning much of West Africa, he gave up his throne, his power and his riches in order to satisfy his curiosity about whether there was a "bank" on the other side of the ocean. In 1312, he set sail from modern-day Gambia, supposedly with a fleet of 2,000 ships—none of which were ever seen again. Some Malian historians argue that he may in fact have reached the shores of Brazil, but even if he did, he never made it back, which let's be honest is a fairly crucial component of the exploration business.

Or perhaps it could never have been different, and this is just the way we are. When you zoom out far enough, a lot of human history is just the story of empires rising and falling and killing the crap out of each other. Like agriculture, and leaders, and war—all the things that in turn helped to kick off the age of empire—they don't necessarily win because they're the best long-term plan for humanity, but because once *somebody* decides to do it, pretty much everybody else has to join in or get crushed. It's like a bar fight in an old Western, except lots of people don't get up when the piano starts playing again.

When Columbus managed to accidentally sink the *Santa María* on the shores of Hispaniola in 1492, the population of the indigenous Taíno people on the island numbered in the hundreds of thousands. A little over two decades later, after the Spanish introduced mining, slavery and disease, there were only 32,000 of them left. Columbus was bad at math, sure, but it definitely wasn't his worst mistake.

It isn't necessarily the job of historians to make moral judgments on the past. They seek to uncover, to describe and to contextualize; to understand and explain how lives long past were lived; and to trace the intertwining webs of power and conflict that gave rise to the world we live in today. You can do all that without passing comment on whether those things were virtuous or wicked. Indeed, given the headache-inducing complexity of it all, it's rarely a simple job to get all judgy about the past.

Fortunately, getting all judgy about the past is *exactly* the job of this book, so let's quickly clarify something: colonialism was bad. Really, really bad.

How bad, exactly? Well, one estimate of the deaths from European colonialism in the twentieth century alone puts the figure in the region of 50 million, placing it up there with the crimes of Hitler, Stalin and Mao—and that's in the century that colonial empires were collapsing. In the hundred years or so following the colonization of the Americas, a fairly conservative estimate is that 90 percent of the continent's population died from a combination of disease, violence and forced labor—again, a figure in the tens of millions. The only reason we can't be more specific is because it's hard to work out how many people were living there before; we literally don't know what we lost.

Of course, the death toll alone, awful as it is in its vagueness, doesn't tell the full story. The African slave trade, the invention of the concentration camp, sexual slavery in the Japanese empire, the Spanish *encomienda* system in the Americas (where conquistadors were personally awarded work gangs of native people, like start-up employees being given human stock options)—the list of horrors is long and almost unbearably grim. And you can add to that the myriad cultures wiped out, the history destroyed and the vast illegitimate transfer of wealth from one part of the world to another, which is still evident in the relative prospects and comfort you're likely to enjoy today depending on which bit of the world you were born in.

Like I said. Bad. This bit of the book isn't very funny, sorry.

All of this probably shouldn't even need saying, but we're currently in the middle of a rather intense Colonialism Was Actually Good backlash, so here we are. The argument, briefly, is that the benefits of colonialism for the colonized and their descendants—the modernization of their economies, the building of infrastructure, the transfer of scientific and medical knowledge, the introduction of the concept of the rule of law—outweigh the regrettable mistakes that were made. But however you dress it up, this still effectively boils down to a claim that the colonized peoples were fundamentally *uncivilized*—incapable of self-governance, immune to progress and insufficiently advanced to utilize their natural resources appropriately. They were just sitting on all this gold, poor idiots, with no idea what to do with it.

For starters, this rests more on myths about the state of precolonial societies than it does on facts, and it inflates a few countries' historically temporary and highly contingent superiority in military technology into some sort of immutable moral law of Who Should Be Allowed to Run Things. What's more, it relies on the unspoken assumption that without colonization, the rest of the world would have simply remained in stasis for the past five centuries, or that there was no conceivable way—other than marching into a country and claiming it as your own—that people could possibly exchange scientific or technical knowledge across borders. Without all that generous colonization, the implication is that they'd still be stuck somewhere in the 1600s. Now that seems unlikely, especially given the transnational exchange of ideas that led to Europe's run of technological advances in the first place, but of course it's impossible to prove one way or the other because there simply aren't enough countries in the world that were neither colonized nor colonizers for us to check. There's Thailand, which almost alone managed to escape. I just googled and it turns out they do actually have electricity in Thailand, so on a sample size of one I suspect this argument might be bollocks.

But ultimately, this is all talking at cross-purposes, because waiting for several hundred years to pass and then doing a sort of retrospective cost–benefit analysis of your actions is not actually how humans generally distinguish right from wrong. That seems more like an after-the-fact attempt to justify what you already want to believe. As a result, the conversation about colonialism tends to involve two people shouting, "But trains!" and "Yes, but also the Amritsar Massacre!" at each other repeatedly, until everybody loses the will to live. (For the record, no, trains are not a moral counterweight to massacres, and I say that as someone who really likes trains.)

None of this involves saying that colonialism is responsible for every ill in the world, which it isn't; or that before the arrival of the colonizers, the societies they would colonize were all blissful oases of peace and comity where everyone lived in harmony with nature, which they weren't. I hope by now in the book it's evident that the capacity for being dumb and awful has been pretty common across world history. It just means that as a species we should probably try to think about our past on the basis of what actually happened, rather than a vague nostalgic yearning for uncomplicated narratives about the glories of empire.

To take just one example: the idea that colonialism brought enlightened governance and enshrined the rule of law in the colonized countries doesn't really square with the history of the numerous treaties signed between colonial powers and indigenous peoples—a history that doesn't exactly scream "respect for the rule of law." That notion would be a surprise to, say, the Native American nations who signed hundreds of treaties with the British and then American governments, only to see every one of them broken and their land taken from them. It would be a surprise to the Maori who signed the Treaty of Waitangi, where a series of translation errors between the English and Maori versions of the text led to some rather convenient ambiguity about exactly what had been signed away. It would be a surprise to the Xhosa people who lived in the southern African colony of British Kaffraria (yes, they literally named the territory after a racial slur for black people), who in 1847 were forced to watch as the newly installed governor, Sir Henry Smith, laughed as he symbolically tore up a peace treaty in front of their eyes, then forced their leaders to come forward one by one and kiss his boots.

Those aren't metaphors by the way. He literally did that. It's worth noting that British history generally remembers Sir Henry Smith as a dashing and heroic figure, one who was immortalized in a popular romantic novel depicting his fairy-tale marriage to [checks notes] a 14-year-old girl.

This all drags us back to one of the themes of this book: our deep and consistent ability to fool ourselves with stories and delusions about what it is we're actually doing. Maintaining an empire requires active and ongoing efforts to mythologize its present and to misremember its past. This dissonance was in place right from the start: it's why Columbus's writing shows he firmly believed he was doing the Lord's work in spreading the Christian faith at the very same moment he was mentally weighing up the Taíno's potential for subjugation and servitude. It's also why the British systematically destroyed tens of thousands of their own colonial records as they left Africa at the end of the imperial age, literally burning them and hurling them into the sea en masse in an effort to erase history and institute a collective amnesia. (In Uganda, this was given the very on-the-nose name of "Operation Legacy.")

And nowhere is it more clear than in the deep, dark irony

of perhaps the most horrifying single act of the colonial era—when King Leopold II of Belgium purchased a million square miles of the Congo Basin as his own personal property, which he turned into a mutilatory, for-profit holocaust of slave labor that resulted in the deaths of perhaps 10 million people over two decades. The irony is this: it was officially done in the name of charity. The land was granted in 1885 to a charitable organization called the International African Association, set up by Leopold. This happened at the Berlin Conference—a meeting in which the countries of Europe carved up Africa between them, catalyzing the "Scramble for Africa" that took colonization of the continent to new extremes. The International African Association's supposed philanthropic mission was to bring "civilization" to the people of the Congo. What it actually did was turn the entire country into an immense rubber plantation where the population was punished for failing to meet production targets by death, or by having their hands or feet or noses cut off. Because the Belgians wanted to make sure their forces weren't wasting expensive bullets on nonessential activities—anything other than killing—soldiers were expected to deliver a requisite number of severed hands to prove how many people they had killed. One bullet, one hand. And so baskets of amputated hands became a kind of currency in the land, one that was harvested freely from both the dead and the living.

Naturally, Leopold called his country the "Congo Free State."

So yes. Colonialism was bad.

This is a book about failure, and while colonialism was definitely bad, it wasn't exactly a *failure*. If you somehow ignore the ethics and just look at the bottom line, it was largely a roaring success, and many of the people behind it made out like kings (particularly the ones who were, in fact, already kings).

But while the big picture is that yes, colonial powers succeeded in getting extremely rich by aggressively stealing the rest of the world's stuff, that misses how an awful lot of the scramble for colonial land was ferociously incompetent. All that self-mythologizing about heroic adventurers, coupled with the lure of supposedly easy money, meant that a lot of the people who threw themselves into the imperial project were—bluntly—complete fucking idiots.

The "Age of Discovery" was rife with the Dunning–Kruger effect writ large. A seemingly endless series of profoundly unqualified, inexperienced and often unhinged men were given expeditions to lead, or colonies to run, on the basis of little more than that they were extremely confident and seemed like the right sort of chap.

Take, for example, John Ledyard, who was entrusted by the British to lead an expedition to find the much-sought-after source of the Niger River, despite his only experience of Africa being a brief nautical stopover on its southern tip. Born in the then-British colony of Connecticut, he developed a reputation as a great explorer thanks to writing a popular book on his voyages as a member of Captain Cook's crew. His solo ventures, however, left something to be desired.

One skill Ledyard undoubtedly had was making friends with important people and persuading them to advance him money. His first venture was a proposed fur trading company, which repeatedly failed to materialize. But while in Paris looking for business partners, he won the backing of various luminaries—including Thomas Jefferson, the Marquis de Lafayette and several others who aren't in the cast of *Hamilton*—for a completely different expedition. This was a bold plan to travel all the way across Russia to the Bering Strait, and from there to cross to Alaska and explore down the entire length of the American continent's west coast. Jefferson, whose idea the whole affair was, described Ledyard as "a man of genius…and of fearless courage and enterprise."

Ledyard lost his shoes on the journey to Saint Petersburg,

but borrowed some money and managed to make it as far as Irkutsk, where the expedition came to an end after he was arrested as a spy.

It was when a penniless Ledyard eventually made it back to London in 1788 that he was given the opportunity to lead the expedition to what was known as "darkest Africa." Despite the fact that he didn't speak Arabic and had at best a patchy track record, the secretary of the African Association—the group who was hiring for the gig—was instantly impressed. The secretary, a Mr. Beaufoy, recounts somewhat breathlessly that on first meeting Ledyard he "was struck with the manliness of his person, the breadth of his chest, the openness of his countenance and the inquietude of his eyes… I asked him when he would set out. 'Tomorrow morning,' was his answer." A single evening might seem a suspiciously short time to prepare for an expedition into uncharted territory on a continent you've only ever seen from a ship, but then, your chest is probably not as manly as John Ledyard's.

In the end, Ledyard made it no farther than Cairo, where he became ill with a "bilious complaint" and tried to self-medicate by swallowing sulfuric acid. Which, not unsurprisingly, killed him. He died in January 1789, the only notable products of his African adventure being a few genuinely useful descriptions of caravan routes, and letters to Thomas Jefferson where he called the Egyptians stupid and slagged off the Nile for not being as good as the Connecticut River.

Or there's Robert O'Hara Burke, an imposing bearded Irish policeman with a furious temper and no sense of direction, who in 1860 set off to explore the center of Australia by tracing a route from Melbourne to the northern coast. Leaving Melbourne to cheering crowds, his party made their way incredibly slowly across the country, largely due to the fact that they were traveling with 20 tons of equipment that included such vital items as a large cedar-topped oak table-and-chair set, a Chinese gong and 12 dandruff brushes.

Thanks to Burke's temper and complete lack of exploring skills, turnover among the expedition party was high, with numerous members either being fired or leaving of their own accord. When their painfully slow progress finally convinced him to dump some supplies, he opted to lose most of their guns and ammunition, plus their supply of the limes that helped to prevent scurvy. Eventually, after around 2,000 miles, having left most of the party behind and taken just three other men and some camels with him, a half-dead Burke made it to within 12 miles of the north coast, before turning back because there was a mangrove swamp in the way. He died on the return journey, shortly after responding to some Aborigines—who'd wandered past and offered the emaciated men food and aid—by firing his gun at them.

Even some technically successful colonial explorers were actually really bad at it. Like René-Robert Cavelier, a Frenchman who ended up claiming much of the American gulf coast for France, and naming what would become the state of Louisiana. Described by one French official as "more capable than anybody else I know," his initial feats of exploration were prompted by his belief he could find a route to China by going through Ohio. He was also an arrogant sod—an unfortunate character trait for an explorer—with a talent for annoying most of the people he traveled with. His final expedition in 1687 was an attempt to invade Mexico and take it from the Spanish with an army of just 200 Frenchmen. After quarreling the whole journey and losing several ships, then missing his planned landing spot by 500 miles, Cavelier was eventually murdered by his own men somewhere in Texas.

But perhaps nowhere is the self-delusion and hubris of the colonial age better illustrated than in the history of the colony that never was—a nation's failed attempt to become a global

Robert O'Hara Burke (1820–1861)

player that instead ended up leaving the country impoverished and humiliated. This is the unhappy tale of the Scottish Empire.

The Man Who Broke Scotland

William Paterson, like many people whose lives have ended up firmly in the "losses" column of history, had a vision.

Not only did he have a vision; he had the skill and tenacity to convince others to go along with it. Paterson was a banker and financier by trade, but a salesman at heart: a man who seemed to combine the rigor of an actuary, the soul of a poet and the fiery conviction of a preacher in one irresistible package. It's just a shame that his particular vision ended up with thousands dead and his home country of Scotland in financial ruin—and worse,

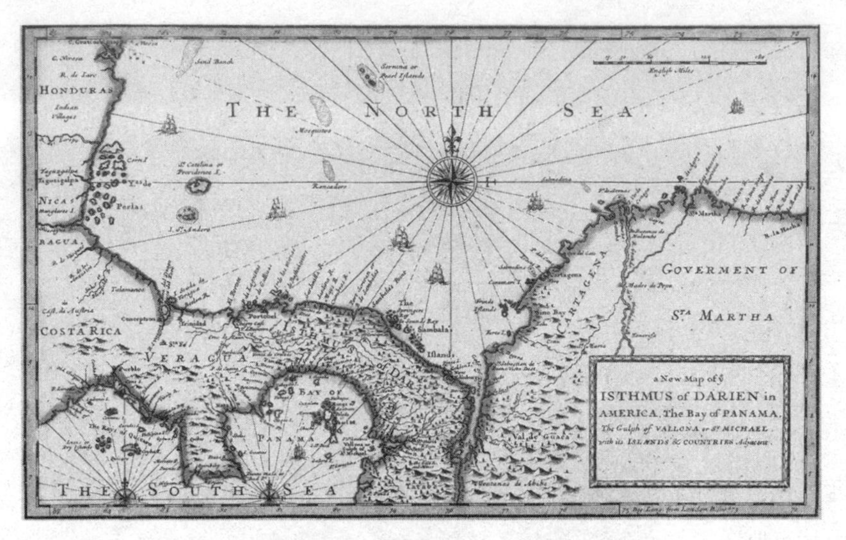

A 1721 map of the isthmus of Darien

at the mercy of its southern neighbor. In fact, without Paterson's disastrous plans, the UK as we know it might not exist today.

It's a story of a country committing itself to grand but vague ambitions based on the proclamations of ideological true believers, of expert warnings not being listened to and of a stubborn refusal to acknowledge reality and change course, even when the world is sending you very clear signals that you might have made a mistake. (It's also a story of the English being dicks, but that probably goes without saying.)

Paterson's vision was of nothing less than a Scottish empire that would become the beating heart of global trade. And he knew exactly where he wanted the first outpost of that empire to be: a verdant paradise on the far side of the Atlantic Ocean, located at the fulcrum of the Americas. That place was called Darien.

Between 1698 and 1699, around 3,000 colonists set sail from Scotland, backed by a wave of nationalist sentiment and as much as half of the country's wealth, giddy with the hope of finding Paterson's paradise and founding that empire. Before the century was out, they'd discovered it was very much not a paradise, most of them were dead and the nation's wealth might as well have been hurled into the waters of the Atlantic.

Now, in fairness to Paterson, not all of his visions were calamitous. In fact, one of his other visions lasts to this day—in 1691, he first proposed, and then in 1694 cofounded, the Bank of England. (And in case you're wondering: a year after the Bank of England was founded by a Scotsman, the Bank of Scotland was founded by an Englishman.) In many ways, Paterson saw far earlier than most how the contours of globalized trade would shape the world we live in today. But he was both optimistic ("Trade is capable of increasing trade," he wrote, "and money of begetting money to the end of the world"), and extremely stubborn. His attitude managed to piss off his fellow Bank of England directors sufficiently that he was forced to resign from the board less than a year after the bank was founded.

And so Paterson returned to the idea that had been something of an obsession for him for many years: a trading colony at Darien, on the eastern coast of the isthmus of Panama, the thin ribbon of land that formed the narrowest point of the American continent. Centuries before the building of the famous canal, it was already clear that Panama was where the journey from the Atlantic to the Pacific and back was easiest. Not *easy*, exactly, as the terrain was far from simple to traverse—but still quicker and safer than the perilous sea journeys via the southern tip of the Americas, around Cape Horn or through the Strait of Magellan. By connecting the two oceans, Paterson wrote with a melodramatic flourish, Darien would become the "door of the seas, and the key of the universe."

This was during the early peak of Europe's wild colonial expansion, and Scotland wanted in on the action. By the 1690s, the Spanish and the Portuguese had been absolutely coining it for the best part of two centuries on the resources they'd extracted from their American colonies; more recently, the English and the Dutch had joined the game to great success. The European scramble for global empires now covered Asia, Africa and the Americas, as the general strategy of "turn up with guns and take all their stuff" continued to promise untold riches, with no sign of slowing down.

The age of empire was also the age of financial revolution: as a result, much of the sharp end of colonialism was enacted not just directly by the states, but also by state-backed, publicly traded "joint-stock" companies that blurred the lines between mercantile business and geopolitics. These included infamous behemoths like the English East India Company and the Dutch East India Company, and it was this model that Paterson sought to broadly replicate for his Darien venture. These companies had a global reach, tremendous wealth and a level of power that outstripped that of many states. Indeed, the companies often acted like states in their own right, and wielded incredible influence over the government of their own countries. (So very unlike today.)

Additionally, the 1690s was also a time of uncertainty and doubt for Scotland. Ever since the Bible-commissioning witch-botherer James VI had gone south in 1603 and united the crowns of Scotland, England and Ireland into one, the Scots had been feeling restive. They were part of a union, yes, but still a politically independent nation: they had their own parliament, passed their own laws and still retained their own currency. However, the suspicion was growing among some segments of Scottish society that they were getting a raw deal from the whole business. The union of the crown, they believed (with some justification) was a stitch-up that acted only in the interests of the English; Scotland would always be the poorer cousin, and the diktats that were passed down from London would always favor the English capital to the detriment of Edinburgh.

These feelings were only increased by the fact that others were actively pushing for an ever closer union with England. And the heightened atmosphere was stoked even further by the financial turmoil of the 1690s—a monetary crisis in England, a king trying to pay for foreign wars and the "seven ill years" of recession, harvest failures and famine in Scotland that saw widespread starvation and the impoverishment of many. This economic crisis, rather than making the people of Scotland risk-averse, instead provided fertile ground for anybody with a promise to shake up the status quo. So when Paterson's Darien scheme came along, it was seized on with patriotic fervor as a way for Scotland to reassert its independence, break free of the binds of the union and take control of its future.

Paterson didn't actually conceive his scheme as a matter of national pride—in fact, he'd been trying to convince other countries to back it before he turned to his homeland. And even once it had become fixed as a Scottish venture in 1695 (as "The Company of Scotland Trading to Africa and the Indies," backed by an act of the Scottish parliament that gave it a wide remit and ridiculously generous terms), he still began his efforts at raising funds for it in London. This is where it all started to go wrong—and where its founders first started to ignore the warning signs.

To begin with, though, things *didn't* go wrong; in fact, they went very right. *Too* right, it turned out. Paterson's reputation in London and his skills as a salesman, added to the unrestrained enthusiasm for joint-stock companies with global ambitions, meant that the Company of Scotland had no trouble finding backers, attracting pledges of investment totaling around £300,000—a vast sum. Unfortunately for them, such was the interest in their scheme that it couldn't fail to attract the attention of the East India Company.

To put it mildly, the East India Company were not wild about the prospect of competition. They, along with much of the rest of London's mercantile community, had been spooked as hell by the financial troubles of the decade, and had recorded massive losses that year. At this point, the Company of Scotland hadn't settled on Panama as their goal and (in the entirely vain hope of keeping things supersecret) hadn't even mentioned the idea of an American expedition publicly. Instead, as the full name of the company suggested, they were selling the scheme as one that would focus on Africa or the East Indies. To which the East India Company's predictable response was, to roughly paraphrase, "Not on your fucking life."

And so the company whose wealth and power was inextricably tied up with the success of the English imperial project put their influence into action. This was the Company of Scotland's first lesson in the brutal realpolitik of global trade: that just because you say, "We want to do lots of international trade," and furthermore that you want to do it on your own wish list of terms, doesn't mean that the rest of the world is simply going to agree with you.

The English parliament was outraged about the terms of the Scottish act, which had shot itself in the foot by granting the Company a free-trade pipe dream: complete exemption from customs and import tariffs and taxes for twenty-one years. How would this affect the customs and trade relationship between England and Scotland, the English MPs wanted to know, and how had the Scottish parliament been allowed to pass it? Lacking a hard border between the countries, they warned that "the said Commodities will unavoidably be brought by the Scotch into England by Stealth…to the great detriment of Your Majesty in Your Customs."

The English parliament held inquiries and ordered reports and threatened to impeach just about anybody who'd been involved in the Company. King William, taking the side of the English (to nobody's great surprise), let it be known that he was royally pissed off. At which point, all of those pledges of investment from London mysteriously vanished into nothingness.

The story was the same when the Company tried to raise funds overseas, in the trading capitals of Amsterdam and Hamburg. The Dutch East India Company was no happier about this state of affairs than their English counterparts, and their efforts—combined with a wily English diplomat who executed a superlative whisper campaign against the venture—ensured that Paterson and his colleagues had lots of coffee meetings where they were pumped for information on their plans, but left with little in the way of actual cash.

But if the efforts of the English state to crush Scottish dreams worked wonders at choking off outside investment, they had precisely the opposite effect inside Scotland. Fueled by the justified sense of unfairness over their treatment, the people of Scotland embraced the Company not just as a financial opportunity but as an expression of national identity. Paterson may not have intended the Darien scheme as a flag-waving exercise—he was only really interested in putting his theories about trade into practice—but ever the salesman, he knew when to ride a wave of public sentiment, and so happily yoked his economic experiment to the surge of patriotic fervor and nationalist resentment.

When the subscription book for the Company was opened in Edinburgh on February 26, 1696, it attracted major crowds, which wasn't exactly normal for what was in effect *Accountancy: Live!* Scots absolutely *poured* money into the scheme. Scotland was not a wealthy country at the time, but even during the seven ill years it was not quite a poor one, either. Like much of the rest of Europe, it had a burgeoning middle class, and they were among the scheme's most enthusiastic backers—unlike other joint-stock companies such as the East India Company, whose investors tended to be limited to the nobility and wealthy merchants. According to historian and author Douglas Watt, who examined the Company's records for his book *The Price of Scotland*, it was small landowners outside the nobility who were the largest single group of backers. But it didn't stop there. A remarkable cross-section of Scottish society pledged their money to the Company—from the titled great and the good to lawyers, doctors, ministers of the church, teachers, tailors, soldiers, watchmakers, at least one "soap-boiler" and even some of the wealthier servants. The enthusiasm was infectious. Tales of the awesome riches waiting in the colonies were the talk of the town; songs and poems were written in praise of the Company, and prayers were said for its good fortune.

It's hard to be precise, given the vagaries of history and the fact that there were two currencies operating in the country, but Watt estimates that somewhere between one sixth and an entire half of *the total monetary wealth in Scotland at the time* was paid into the Company's coffers. When you include the full amount pledged (as only part of the cash was required to be paid up front), it's possible that the promised funds actually exceeded the total value of coins in the country.

This is, just to be clear, not a good thing.

Paterson seems to have understood well how financial manias could be fueled, and used this to his advantage. In fact, he discussed it in terms that seem eerily like our modern understanding of "going viral." He wrote in a 1695 letter that "if a thing goe not on with the first heat, the raising of a Fund seldom or never succeeds, the multitude being commonly ledd more by example than Reason." One key factor may have been that the subscription book for the Company was not private, but public, and indeed was deliberately published by the Company so everybody could see who the investors were. And Paterson deliberately targeted prominent public figures ("influencers," if you will) to be early supporters, in the hope that they would be the example to others that would lead them more powerfully than reason. Like some sort of seventeenth-century Kickstarter or GoFundMe page, this turned the act of backing the Company from a personal financial choice into a public declaration of allegiance—and it made those *not* backing it conspicuous by their absence.

Naturally this all led to a self-reinforcing spiral of social pressure, and created an atmosphere in which opposing or skeptical voices were aggressively drowned out. In 1696, John Holland (the Englishman who founded the Bank of Scotland) recorded unhappily that when he tried to criticize the scheme, he was accused of being a spy for the East India Company. "Such is the zeal of the nation to the Indian and African Trade," he wrote, "that many are thereby prejudiced against me; and because they cannot answer what I have argued against their design, they tell one another, we must not believe what Mr. Holland sayeth, for he is an English man...it is become dangerous for a man to express his thoughts freely of this matter, people being under more awe and fear of giving their opinion..."

The combination of outrage over English actions, surging patriotic self-belief, lofty promises and a compelling vision, the trick of turning support into a performative act and the good old-fashioned lure of making a quick buck created just about the most fertile possible environment for a runaway mania. And so it was that on July 14, 1698, as cheering crowds waved them off, five vessels set sail from Leith, carrying aboard them William Paterson and 1,200 other hopeful souls, all bound for a Central American destination that Paterson had never been to.

Oh yeah, had we not mentioned that bit? WILLIAM PATERSON HAD NEVER EVEN BEEN TO DARIEN.

Quite why our boy became so fixated on Darien as the site of his grand trade experiment remains to this day a bit of a mystery. He had certainly spent a lot of time in the Caribbean as a merchant, but there's no evidence in his biography or his public writings that he ever came anywhere close to the Panamanian isthmus. Instead, he seems to have heard tales of it from, in all probability, pirates. (This was during the Golden Age of Piracy, when the real-life, non-CGI Pirates of the Caribbean were doing their thing, either as true rogue elements or often with the nod-and-a-wink backing of governments who wanted them to harass their colonial rivals.)

It's also not clear quite how Paterson so consistently persuaded his fellow directors of the Company of Scotland to back his vision of Darien as the hub of Scotland's global empire, based on little more than hearsay. Certainly they had plenty of opportunities to change course—in 1697, a year before the fleet set sail, they actually came close to abandoning the Darien scheme entirely and focusing instead on more modest goals.

They were becoming aware that the Company, flush with cash after its fundraising in Edinburgh, had now wildly overspent and could not guarantee funds to fully support the scheme's ambitions. (They'd foolishly decided to purchase entirely new state-of-the-art ships on the continent, at a time when most of their rivals rented the majority of their fleet. Possibly this was an effort to big themselves up to potential Dutch and Germanic investors—kind of like a tech start-up with no revenue but swanky offices in the most expensive part of town.) The directors had multiple experts of good standing casting doubt on the viability of the expedition, urging instead that the capital raised be spent on less imperial trade missions to Asia. They were fully aware of all the pitfalls of Darien as a destination, and even considered several other locations in the Americas that might have been better suited…yet still this group of sober, well-educated and terribly respectable individuals convinced themselves that they'd been right all along, and decided to forge ahead.

Exactly what those pitfalls were started to become clear shortly after the colonists arrived, at the beginning of November 1698. Many of them hadn't even been aware that Darien was their destination: their orders were only revealed once the ships had sailed, as part of the Company's hopeless efforts to keep their plans secret from rivals.

To begin with, things seemed pretty great. The settlers were awestruck by the location's natural beauty and the (to them) alien species—the land turtles and sloths and giant anteaters. The local Guna people seemed friendly, and spoke of gold mines just a few miles away. The settlers were delighted to discover a "most excellent harbor," a naturally sheltered two-mile-long bay that one of them, Hugh Rose, believed was "capable of containing 1,000 of the best ships in the world." Another anonymous diarist wrote that "the Soil is rich, the Air is good and temperate, and everything contributes to make it healthful and convenient."

"Healthful" may have been an overstatement. Before long, some of the colonists began to get sick and die. William Paterson's wife was one of the first to pass, less than two weeks after they landed. A few days after that, the colony's last religious minister perished, as well.

But despite these tragedies, the settlers remained confident. They named the bay Caledonia, after the old name of Scotland, and immediately set about constructing their first town, which they called New Edinburgh. So delighted were they with their finds that they dispatched the expedition's chief accountant, Alexander Hamilton (not the one from the musical), to make the return journey on a passing French pirate ship and deliver the glad tidings to home.

A fairly clear sign of how badly wrong things were actually going came when Hamilton's ship sank as soon as it left the harbor.

At this point, it became apparent exactly why such a large natural harbor was going unused by any other colonial power. Rather like the Hotel California, getting into it was a doddle, but leaving was quite another matter. The prevailing winds blew in such a way that upon exiting the shelter of the bay, ships were immediately forced backward and assailed by huge waves. The ship carrying Hamilton was smashed to pieces in around thirty minutes, drowning almost half the crew. (Hamilton himself survived, and would eventually make it back to Scotland to tell everybody how well the expedition was going.) The Company had been warned by experienced sailors that their large, expensive ships with shallow keels were entirely unsuited to the conditions of the Caribbean, but they'd ignored that advice. For a proposed trading colony, you'd think the fact that their ships were going to be stuck in the harbor for many months of the year might have prompted second thoughts, but no.

It's also questionable how clearly they'd thought the whole trade thing through. For a trade mission, Douglas Watt's research suggests they'd spent a remarkably small amount of their budget on tradable goods—which mostly consisted of major stores of cloth, but also included over 200 periwigs, a sizable stock of fashionable shoes and a large number of combs. (The last were possibly brought along in the belief that native peoples around the world universally lose their shit at the sight of a comb, and will promptly trade away their land. In the end, the Guna seem to have given not the slightest toss about combs.) On the other hand, if the mission's goal was simply to establish a settlement, they could maybe have done with slightly fewer wigs and a couple more tools instead.

As the task of building New Edinburgh began, morale quickly began to plummet. The work was backbreaking, and taking place in entirely un-Scottish heat. When, after two months of fruitless hacking away at thick jungle that never seemed to yield, the project leaders decided they'd been building in the wrong place all along ("a mere Morass," as Paterson described it), spirits sank even lower. Then the rains began—and the rain in Panama is not like the rain in Scotland. The diarist Rose also changed his positive opinion of the location rather quickly, now writing: "On the main and all the bay round full of mangrow and swampy ground, which is very unwholesome."

The swamps were more than unwholesome. The sickness that had already killed Paterson's wife started to take the colonists in ever larger numbers. It's not clear what it was, as they merely recorded it as "the fever," but the best bet is malaria or yellow fever thanks to mosquitoes in those nearby swamps. (Both diseases, of course, were themselves colonists, having been helpfully brought over from the Old World by Europeans.) The settlers were dying at an alarming rate.

Those who didn't get sick from the fever were increasingly destroying their health in other ways, thanks to the fact that one major perk of the trip the Company of Scotland had decided on was a plentiful supply of liquor. The people of Caledonia started to drown their sorrows in rum and brandy, which did not make the work of building New Edinburgh go any quicker. After a while, the leaders decided to abandon construction of the town entirely and focus on the creation of a fort, as they grew increasingly wary of a large-scale Spanish attack.

Ah yes. The Spanish. You see, we haven't actually mentioned the biggest, most startlingly obvious problem with Paterson's scheme yet: the fact that the Spanish were pretty bloody sure they already owned Darien.

They got this notion from a few little things. Like them having been active on the Panamanian isthmus for almost two centuries. Like how it was a vital route for them to ship their plundered South American gold and silver back to Spain. And like the way that Darien lay right between three of their major cities. They had actually occupied Darien in the past, before abandoning it due to all the problems the Scots were just now discovering. The idea that Spain was about to let an upstart country simply waltz in and establish a new colony slap bang in the middle of their own was laughable.

How did the Company of Scotland ever think the Spaniards would let them get away with this? It's a true head-scratcher. But here we do at least have a general idea of their thinking. Buoyed by romantic pirate tales of successful attacks on Spanish properties in the area, they seem to have believed that Spain had become a paper tiger, a fading imperial power with its best days long behind it. Despite the fact that Spain's navy outnumbered Scotland's (by one navy to zero), they likely believed that if they could repel any initial attacks, they'd be able to successfully call their opponent's bluff.

That's…not quite how it played out. For starters, Spain didn't have to attack directly. If the English attempts to thwart Scottish ambitions had been damaging beforehand, it was nothing compared to what happened now. The Spanish swiftly and diplomatically let King William know that Scotland's little venture was exactly the sort of shit wars get started over. Having only just extracted himself from one of England's regularly scheduled wars with France, William was desperate to maintain peace with Spain, and so he immediately issued orders that no English territory or ship was to supply, give aid to or even correspond with the Scots in any form.

When news of this reached Caledonia, it sent the settlers into despair. They'd had no news from home since they arrived, no fresh supplies had come despite regular pleas sent back to Scotland—and now they were entirely cut off, with any hope of finding allies in the region dashed.

Before the English embargo even came down, the colonists had already fought off one small Spanish strike, which they'd been warned of in advance by the captain of an English ship sent to spy on their activities. (Humiliatingly, he'd actually arrived in the area before them, because the Company's attempts at secrecy had been so inept.) That small victory improved morale for a while, but was canceled out when one of their ships was seized by the Spanish while it was out looking for people to trade with, its crew thrown in jail and its cargo seized.

Now, with half the population of Caledonia dead, dying or imprisoned and the other half exhausted, starving and hungover, the news that they were completely isolated was the last straw. Believing themselves entirely abandoned, en masse they opted to abandon Darien and make the sad journey home.

And so a mere nine months after William Paterson had finally arrived in the place he'd dreamed of for much of his life, widowed and now sick himself, he was carried on board a ship preparing to leave it. He survived the fever, but he would never see Darien again.

Still ravaged by sickness, the colonists' journey home via Jamaica and New York was as awful as their time on Darien had been. It took them almost a week just to get out of the harbor, and hundreds more died en route. One ship sank, and another was nearly destroyed. In the end, only a single ship would limp all the way back to Scotland. Where, unfortunately, it didn't arrive in time to prevent a second fleet sailing to Darien to find out what had happened to them.

That's right, the Company of Scotland had finally decided to send the long-overdue reinforcements, just when it was too late.

This second fleet arrived at the end of November 1699 to find a "howling wilderness": the abandoned and burned-out remains of New Edinburgh, an overgrown fort and a large number of shallow graves. Against all reason, the new arrivals decided to stay, rebuild and try to hold on to the land while sending for fresh supplies. All this actually achieved was yet more of them becoming sick and dying, and gifting Spain the opportunity to prove they were no faded power. A few months into the new century, the Spanish arrived in force to remind everyone who was still boss. Ravaged by fever, the Scots somehow managed to hold out under siege for a while, but by April they were forced to surrender. The Scottish Empire was over.

Possibly understanding the propaganda value of a defeated enemy fleeing with their tails between their legs, or maybe just feeling sorry for the poor bastards, the Spanish allowed the settlers to go. Once again, hundreds died of fever on the journey back. A violent storm destroyed a further two ships with the loss of a hundred more souls—including the remarkably unlucky accountant Alexander Hamilton, who having made it back to Scotland despite his first shipwreck had then opted to return to Darien with the second fleet.

In total, somewhere near 3,000 people had sailed from Scotland for Darien. Between 1,500 and 2,000 are thought to have died either in the bay of Caledonia or on the seas. Many of the survivors never returned to Scotland.

Back in Edinburgh, the failure of the scheme prompted waves of shock as the news trickled back over the course of 1700. In a newly polarized political environment, the issue became a political football, with reaction split between those blaming the Company's directors for their shameful failure, and those blaming the perfidious English for their interference. There were riots in Edinburgh in support of the Company. One disgruntled colonist whose pamphlets tore into the directors of the Company was accused of blasphemy; three Company supporters who produced a derogatory engraving attacking the government were unsuccessfully tried for treason. It didn't really matter what the facts were anymore—it was all about which side you were on.

The effect was not merely political, but financial: in the midst of an economic crisis, a significant percentage of the country's total wealth had been thrown away. The individual investors had lost large sums with seemingly no hope of return. Scotland had been humiliated and weakened.

Of course, no major political change happens only for one reason. The forces pushing Scotland toward a full union with England were complex, and didn't just spring into existence in the wake of Paterson's foolhardy scheming. This was the end of the seventeenth century, after all, when borders and alliances seemed to change every other week. But Darien sure as hell contributed to it—particularly when, as part of the union deal a few years later, it turned out that England was offering Scotland a bailout. Not just for the country; for the individual investors in the Company of Scotland, who would receive their original stake back with a generous sum of interest.

Many called it a bribe. "We're bought and sold for English gold," as Burns would write eight decades later. Some saw the whole affair as a dark English plot to cripple Scotland to the point where it would have no options left. Others were just happy to have their money back.

Paterson argued in favor of union.

In May 1707, the United Kingdom came into existence. In August, a dozen heavily guarded wagons containing almost £400,000 rolled into Edinburgh.

The thing about all this is: Paterson wasn't wrong, not exactly. Panama really was an excellent site for a colony—indeed, archaeologist Mark Horton surveyed the isthmus in 2007 and concluded that Paterson's proposed trade routes from Darien were actually realistic. And his forecast of how global trade would develop doesn't sound so far off the mark today, either; what's more, he explicitly promoted it as a nonviolent alternative to the atrocities of empire, writing that trade could bring wealth without "contracting such guilt and blood as Alexander and Caesar." Which frankly for the time makes him almost kind of woke. (Although let's not go overboard: gleeful talk of Darien's untapped gold mines indicates that plenty of the scheme's backers were in this for the plunder of natural resources.)

What really doomed the venture was a collective failure on the part of the scheme's backers to grapple with difficult questions. They brushed aside the details, such as the type of ships they'd need and what supplies to take; they simply ignored the big picture, like the geopolitical implications of their actions. Instead, when setbacks or pitfalls emerged, they ended up believing their own hype and convincing themselves ever more strongly that they'd been right all along. It was a classic case of groupthink.

To this day, the story of Darien is one that divides Scotland. During the 2014 referendum on independence, it became a metaphor for both sides. For the nationalists, a parable of how England had always sought to sabotage and oppress Scottish hopes;

for the unionists, a lesson in the dangers of abandoning stability in favor of unrealistic ambitions.

As a tale, it lends itself to metaphor. I mean, it's the story of a country turning away from a political union with its closest geographical trading partners in favor of a fantasy vision of unfettered global influence promoted by free-trade zealots with dreams of empire, who wrapped their vague plans in the rhetoric of aggrieved patriotism while consistently ignoring expert warnings about the practical reality of the situation.

Unfortunately, I can't think of anything that could be a metaphor for right now.

5 MORE EXPLORERS WHO FAILED AT EXPLORING

Louis-Antoine de Bougainville

A French explorer who, while becoming the first Frenchman to circumnavigate the globe, got as far as the Great Barrier Reef but turned back, thus failing to discover Australia.

John Evans

A Welsh explorer who spent five years in the 1790s searching for a lost Welsh tribe in America, during which he was imprisoned as a spy by the Spanish, before eventually finding the tribe—the Mandan—and discovering they weren't Welsh.

Vilhjalmur Stefansson

A Canadian explorer who believed that the Arctic was actually a pretty hospitable place, and led an expedition there in 1913. When his ship got stuck in the ice, he told his men he was leading a small party to find food, and then promptly abandoned them.

Lewis Lasseter

In 1930, Lasseter led a search party into the central Australian desert in search of a vast "reef" made of pure gold that he claimed to have found years before. There's no such thing. Eventually the rest of his party abandoned him, then his camels ran away while he was doing a poo, and he died.

S. A. Andrée

A Swedish engineer and adventurer, Salomon Andrée came up with the excellent idea of reaching the North Pole by hydrogen balloon—and set off despite the fact that the balloon was leaking gas. He and his crew died somewhere in the Arctic.

8

A DUMMIES' AND/OR CURRENT PRESIDENTS' GUIDE TO DIPLOMACY

As global travel exploded in the Age of Discovery, so, too, did the opportunities for accidentally starting all manner of wars, as it dramatically increased the number of countries you could infuriate. On the assumption that, at least sometimes, you do actually want to avoid having wars, then (short of doing whatever the unclear thing was that the Harappan people did) your best bet is diplomacy. Diplomacy is the art of large groups of humans not being wankers to each other—or at the very least, managing to agree that okay, everybody is a wanker sometimes, but why don't we try to take it down a notch.

Unfortunately, we're not very good at that, either.

The key problem with international relations stems from a more general and fundamental problem of human interactions, namely that it involves two basic principles:

1) It is a good idea to trust people.
2) But not too much!

This is the dilemma that haunts pretty much every moment of contact between different cultures in history. Unfortunately for the people living in those moments, there's no way of knowing which choice will be the right one. That's a problem we've still not quite figured out, but at least we have the luxury of looking back at people's choices in the past and going, "Nope, definitely the wrong call."

It's the problem that the Taíno faced when Columbus came along—during their first meetings, they were trusting, and impressed Columbus with their friendliness and generosity. Obviously Columbus reacted in the normal way you do when someone is friendly and generous to you: "They should make good servants," he mused, adding after thinking about it for a few more days that "with fifty men they can all be subjugated and made to do what is required of them." Lovely chap.

Roughly the same thing played out on a grander scale some decades later, when the Aztec ruler Moctezuma made a very, very bad decision about the intentions of Hernán Cortés.

The Aztecs (or Mexica, as they called themselves) ruled a large empire that stretched from coast to coast across what is now central Mexico. Moctezuma led it from the city-state of Tenochtitlan, the largest and most advanced city on the continent (it's where Mexico City stands today). Everything was going pretty well for them until 1519, when Cortés landed on the Yucatán coast.

Cortés was not just a conquistador, but a rogue conquistador—he'd actually been removed of his command of the exploratory mission by the Spanish governor of Cuba, who didn't trust him, but Cortés just took the boats and the crew and went, anyway. A while after his arrival, he deliberately sank the boats to stop his crew mutinying and heading back to Cuba. What I'm saying here is, Hernán Cortés was not a team player. And at this point, on the run from his own countrymen and with no way to get home, he'd pretty much used up any options that weren't "conquer stuff."

When Moctezuma heard of Cortés's arrival, some 200 miles from Tenochtitlan, he was understandably nervous. Unfortunately, he couldn't decide what to do. He vacillated between sending Cortés lavish gifts and sending him warnings to stay the hell away. Cortés, meanwhile, was busy exploiting the Mexica's weaknesses. The main problem was: they were an empire, too, and an often quite brutal one. As such, there were plenty of native groups in Mexico who weren't big fans of Moctezuma, and as Cortés made his way inland he used a combination of smooth talking, trickery and occasional mass slaughter to persuade them to ally with him against Tenochtitlan.

All this should probably have been a sign to Moctezuma that things were probably not going to end in a new era of friendship, but still he waited. It's possible that his uncertainty was increased by the supposedly widespread belief that Cortés might be the returned incarnation of the sky god Quetzalcoatl—although the only actual evidence that anybody believed this is that Cortés talks a lot about it in his letters, and frankly it sounds like the sort of bullshit he'd say.

When Cortés finally arrived at Tenochtitlan, accompanied by a few hundred Spanish soldiers and a load of his new allies, Moctezuma finally made his decision, despite lots of advisers telling him this was a really bad idea. In fairness, it's not clear if there was a *right* decision he could have made, but this was definitely the wrong one: he invited the Spanish in as honored guests. He showered them with gifts, he gave them the best rooms, the works. This did not end well. Within a couple of weeks, Cortés staged a coup, took Moctezuma hostage in his own court and forced him to rule as a puppet. The first thing the Spanish demanded was dinner; after that, they promptly insisted he tell them where all the gold was kept.

It all blew up a few months into 1520, ironically while Cortés was away fighting a large regiment of Spanish troops who'd been sent by the governor of Cuba to try and stop whatever the hell it was he was doing. One of Cortés's lieutenants who'd been left behind to keep Tenochtitlan under the thumb decided, for no clear reason, to massacre a large number of Mexica nobles in the Great Temple while they were celebrating a religious festival. Outraged at the slaughter, the Mexica people revolted, and Cortés returned to face an uprising. He ordered Moctezuma to tell his people to cease hostilities. They didn't, and that was the end for Moctezuma. Spanish accounts say that he was stoned to death by an angry crowd of his own people; in all likelihood, he was actually murdered by the Spanish when it became clear he was no longer any use as a puppet. Just over a year of bloody fighting later, the Spanish had entirely conquered the Mexica, and Cortés—suddenly back in favor with his bosses—was made the governor of Mexico.

Possibly nobody could have stopped the Spanish invasion, but Moctezuma's decision to welcome them as guests has to go down as one of the most ill-advised pieces of international relations policy in history. And to be honest, if the Mexican government had pondered his example 300 years later when they started encouraging American immigration into Texas, then the key lesson of Moctezuma's sorry tale—"for God's sake, Mexico, stop inviting white people into places"—might have led to history playing out very differently.

Fortunately for Moctezuma's reputation, he's not alone in history's hit parade of poor international relations choices.

The crucial nature of choosing your friends wisely can be

seen in the story of the Roman governor of Germany in 9 CE, Publius Quinctilius Varus. Varus was trying to do the classic occupying-force thing: picking and choosing local nobles to be on your side, in order to keep the peasants relatively placid. Unfortunately for him, he chose to put his trust in a Germanic tribal leader named Arminius, on the grounds that he had been made a Roman citizen and even led an auxiliary unit in the Roman army. Despite being warned that his trusted adviser might not be quite on the straight and narrow, he chose to believe Arminius when he told him there was an uprising among German tribes that needed to be put down. Arminius steered Varus and his Roman legions right into an ambush—one which he himself led, after pulling the old "I'm just going to ride ahead to check things out" trick. Three whole Roman legions were wiped out (the worst military defeat in their history) and the Roman Empire's northward expansion was stopped in its tracks.

On the flip side to being overly trusting, there's the self-defeating Chinese foreign policy under the Ming Dynasty, which has become a case study in the perils of isolationism. In the first three decades of the 1400s, China had one of the greatest naval fleets in the history of the world, under the command of the legendary mariner Zheng He. Composed of up to 300 ships, including enormous nine-masted vessels larger than any boat that would sail for centuries in the future, the fleet carried as many as 30,000 men; it even included ships that acted like floating farms, growing vegetables and keeping herds of animals.

What's more, during this period the Chinese were notable for not really using their fleet to do much in the way of, you know, invading places. Certainly they spent quite a lot of time fighting pirates, and the fleet was very handy for vaguely threatening displays against any country considering stepping out of line—but in all of Zheng He's seven voyages to destinations across Asia, Arabia and Eastern Africa, it was only involved in one relatively minor war. Instead, it spent most of its time visiting ports as far-flung as Malacca, Muscat and Mogadishu, and... well, exchanging presents. The Chinese gave out precious metals and fine cloths, and would get a wide variety of gifts back, including an awful lot of animals. One time they brought a giraffe back from Kenya.

As displays of overwhelming force by imperial powers go, it all sounds rather nice compared to the alternatives. And so it's particularly baffling that after Zheng He's death in 1433, the Ming Dynasty just sort of...stopped. They abandoned their navy. In a pretty extreme overreaction to the continued operation of Japanese pirates, they resurrected the old policy of *haijin*—an almost complete ban on any maritime shipping. Distracted by ongoing battles with the Mongols in the north, foreign diplomatic missions were seen as an unnecessary expense, the money better spent on a different project: the construction of a very large border wall.

In the following years, China turned increasingly inward, shutting out the world. The fact that they did this just at the point when European navies were starting to explore the globe had a double effect: it meant that when the Europeans started to show up in Asian waters a few decades later, there wasn't a major local power to stand in their way, and it meant that China kind of missed out on much of the scientific and technological acceleration that was kicking off. It would be a long, long time before the country regained its status as a world power.

That highlights just how much diplomatic choices are, to an extent, about trying to predict how power balances will shift in the future. Given that this is impossible to do with anything close to accuracy, it's not entirely surprising that people get it wrong so often. In Switzerland during the late spring of 1917, right in the midst of World War I, a middle-aged man with a funny beard had a proposal for the German government. He was Russian, and he desperately wanted to get back home to his own country, which was in the grip of political upheaval, but the war made traveling across Europe all but impossible.

The best route back to Russia was to head northward through Germany, but the man would need German permission to do that. And the German government were no fans of his politics.

The pitch was simple. For all their differences, he and the Germans currently had a shared enemy: the Russian government, who he didn't like and was rather keen to overthrow. The German High Command were fighting a war on several fronts, and reasoned that any distraction that might divert Russian resources away from the front lines would be helpful. So they agreed. They put the man, his wife and 30 more of his compatriots on a train to a northern port, from where they'd travel onward via Sweden and Finland. It wasn't much of a vanguard, but it was better than nothing. The German authorities even gave them some money, and would continue to help them out financially over the following months. They probably imagined that like most political obsessives with a niche cause, the man would stir up a bit of trouble, get the Russians off their backs for a while and then fade quietly into obscurity.

Anyway, yeah, that guy was Lenin.

Now in many ways, the German plan worked flawlessly. Better than expected, in fact! The Bolsheviks didn't just irritate and distract the Russian authorities, they absolutely wiped the floor with them. In just over six months, Russia's provisional government was gone, Lenin was in power and the Soviet state was established. The Germans got a ceasefire that they couldn't even have dared to dream about back in April when they'd waved Lenin's train off.

In the slightly longer term, however, the plan was not what you'd call a runaway success.

For starters, the ceasefire on the Eastern Front didn't actually help win the Germans the war. And subsequently, the relationship between the expansionist new Soviet state and their helpful German pals rapidly turned sour. Fast-forward a couple of decades and another world war later, and half of a newly divided Germany would be under Soviet control.

Vladimir Ilyich Ulyanov, better known by the alias Lenin (1870–1924)

The Germans had fallen into the old trap of believing that their enemy's enemy would be their friend. Which isn't always wrong, exactly—it's just that the friendship usually has a remarkably short shelf life. And that enemy's enemy delusion lurks somewhere in the background of an astonishing number of history's worst decisions, in addition to explaining several centuries' worth of extremely confusing European history.

Another name for this phenomenon could be "postwar US foreign policy." During the extended period of worldwide poor decision-making that was the Cold War, the US allied themselves with just about anybody who met the exacting criteria of "not being a communist." Many of these allies were simply flat-out bastards (see: sundry dictators in Latin America, the succession of awful rulers in Vietnam). But on top of that core problem comes another one: these allies frequently had a habit of turning out to never have been huge fans of the US in the first place.

Consider that, in only the last few decades, the US has been involved in an armed conflict against al-Qaeda, which emerged from the mujahideen in Afghanistan, a group the US had previously supported on the grounds that they were fighting the Soviets. (I strongly recommend watching the 1987 James Bond film *The Living Daylights* if you're a big fan of shouting, "OH WOW THAT DID NOT AGE WELL" at your screen. Bond teams up with the mujahideen, who are led by a charming, heroic character best described as "suave bin Laden with a posh English accent." Good theme song, though.)

In that time, the US have also been involved in an armed conflict against Iraq, a country they had previously supported because they were fighting Iran, a country that opposed the US on the grounds that the US had supported Iran's previous dictatorship, because they, too, opposed the Soviets.

And they've been in an armed conflict against ISIS, who grew out of the activities of al-Qaeda in postwar Iraq and are now fighting in Syria in what is *at minimum* a three-sided war, in which the US opposes a regime it previously supported and then tried to support the enemies of, but it turned out that some of the enemy regime's enemies were also friends with ISIS, who are the enemies of both the US and the US's enemies, although some other friends are enemies of both—oh, and Russia's fighting there, too, just for old times' sake.

And that's just in one part of the world.

Look, international politics is really hard. There's not much room for lofty ideals, and the cold hand of pragmatism means you often have to make do with the allies you can get, rather than the allies you really want. But a lot of the problems we run into time and time again might be avoidable if we remembered that, most of the time, the enemy of our enemy is just as much of a bastard as the original enemy.

But in the long history of really bad diplomatic mistakes, there's one that stands out above all others.

How to Lose an Empire (Without Really Trying)

In 1217, Ala ad-Din Muhammad II, the shah of the vast and powerful Khwarezmian Empire, received a neighborly message from the leader of a new power that had been growing in the east. "I am master of the lands of the rising sun," it said, "while you rule those of the setting sun. Let us conclude a firm treaty of friendship and peace." It proposed a trade deal between the two powers, to their great mutual benefit.

At which point Shah Muhammad II made the absolute worst decision in the long history of international diplomacy.

The Khwarezmian Empire was one of the most important in the world at the time, stretching almost from the Black Sea in the west to the mountains of the Hindu Kush in the east, from the Persian Gulf in the south to the Kazakh Steppe in the north. It covered a huge area that today includes either all or a large part of the territories of Iran, Uzbekistan, Turkmenistan, Tajikistan, Azerbaijan, Afghanistan and more. At a time when Europe was still a century or two away from getting its Renaissance on, Khwarezm was right at the epicenter of the developed world. It was through Khwarezm that the Silk Road ran, the great route that connected east and west, along which both goods and ideas flowed. The shah's domain was one of the beating hearts of the Islamic world, by far the richest and most advanced culture in existence. Cities like Samarkand, Bukhara and Merv, the jewels of the Khwarezmian Empire, were among the great cities of Central Asia, renowned as places of scholarship, innovation and culture.

If you're thinking: that's odd, I've never even heard of the Khwarezmian Empire—yeah, *there's a reason for that.*

You see, the message the shah had received was from a guy called Genghis Khan. And just a couple of years after he made his terrible decision...well, there wasn't a Khwarezmian Empire anymore.

Genghis Khan depicted in battle, from a fourteenth-century book by Rashid-al-Din

It's worth noting that as far as history can judge, Genghis's message of friendship to the shah was completely sincere. By this point, the great warrior had effectively achieved all his goals: he had conquered and united the nomadic peoples of northern China and the surrounding territories into his Mongol Empire, a series of conquests that ranged from the relatively easy to the profoundly brutal. He still had a few battles to win in the east, but no plans at all to push any farther west. This was as far as his ambition and desires stretched; and furthermore, he was nearing his sixties. Job done, time to start planning for a quiet retirement.

It was his recent conquest of the Qara Khitai—an empire of displaced Chinese nomads centered roughly on modern Kyrgyzstan, and one of the last holdouts against his reign—that had brought Genghis to Khwarezm's doorstep and created a new border between the Mongol and Islamic worlds. As tends to happen at borders, especially ill-defined ones, there had already been one abortive military skirmish between the Mongol troops and the Khwarezmians. This took place when Muhammad II and his army turned up to have a battle with some of his enemies, only to find that the Mongols had annoyingly got there first and routed them already.

This wasn't even the first time this had happened. Genghis seemed to have a habit of turning up first and winning wars that Muhammad had been planning on having himself—which might possibly help explain the shah's ill-advised reaction to the olive branch that Genghis extended after that initial skirmish. He was, possibly, a little pissed off and humiliated that the Mongols kept on stealing his glory. (It should also have probably tipped him off that they were pretty good military tacticians, but eh, apparently not.)

Additionally, Khwarezm–Mongol relations seem to have suffered the kind of problems you always get when things are lost in translation. "I am master of the lands of the rising sun while you rule those of the setting sun" was probably just Genghis laying out some basic east–west geography and acknowledging their status as (roughly) equals. But here's an alternative translation of the message: "I am the sovereign of the sunrise and you are the sovereign of the sunset." Put it like that, and all of a sudden it sounds like Genghis is throwing a hefty amount of shade. To a ruler who was already feeling a little touchy about somebody else winning his battles for him, did it come across as effectively saying, "I'm a rising power and you're a fading power, LOL"?

Conducted via a series of emissaries sent back and forth, the subsequent dialogue between Muhammad and Genghis plays out like a passive-aggressive comedy of manners. Genghis felt patronized by the gifts of fine silks the shah sent him ("Does this man imagine we have never seen stuff like this?"). He responded by sending back a gift of an enormous gold nugget, presumably in an effort to demonstrate that they had nice things, too, even if they lived in tents. At which point Genghis's earnest repetition of his wish for peace—"I have the greatest desire to live in peace with you. I shall look on you as my son"—hit entirely the wrong note with Muhammad, who really was not a fan of being called "my son." (This is much funnier if you say "my son" in a gruff cockney gangster voice, by the way.)

And yet, with formalities and protocol still being observed (despite the undercurrent of pettiness), Genghis clearly believed that his request for a peaceful trading relationship had been agreed to. For starters, it was so obviously a win-win for everyone. "You know that my country is an ant heap of warriors, a mine of silver, and that I have no need to covet other dominions," as he told Muhammad in one message. "We have an equal interest in fostering trade between our subjects."

And so it was that Genghis sent out his first trade mission to Khwarezm, backed with his own funds and led by his personal envoy: 450 merchants, 100 troops and 500 camels, with wagons loaded with silver and silk and jade. Their aim was, in the first place, to ensure that Khwarezm's recent embargo on trade across the border with the Mongol Empire was at an end. Everybody was extremely keen for this to happen, especially beyond Khwarezm's borders: Genghis's unification of northern China had made passage along the Silk Road theoretically much easier, and merchants across the Islamic world were superkeen on a chance to crack the Chinese market. But the shah's territorial pissiness had closed the route off. As such, when the caravan of merchants and goods entered the northern Khwarezmian city of Otrar in 1218, it must have seemed like good times were here again.

This is where it all went very, very wrong, for a very large part of the world.

Instead of welcoming the trade mission, letting them park their camels and offering them a nice cup of tea, Inalchuq Qayir-Khan, the governor of Otrar, took a different approach. He had them all killed and stole everything they had brought with them. It was a vicious surprise attack, in which only one person out of the 550 in the party survived, because he was having a bath at the time of the massacre and managed to hide behind the tub.

The incident shocked the world, as an outrage against decency and hospitality and also basic common sense. The explanation given by Inalchuq—that he suspected the entire party of being spies—was completely ludicrous. The merchants themselves weren't even Mongols, instead being largely Muslims from the Uighur region. The prospect that Islamic merchants in an Islamic city on a major trade route were now at risk of being massacred by the local government on a flimsy pretext was, to put it mildly, rather upsetting and definitely not good for business.

And absolutely nobody believed that Inalchuq would do something so potentially destructive—for an empire whose wealth and prestige depended on trade—without either the permission of, or a direct order from, the shah himself.

If there were any lingering doubts that Muhammad was hell-bent on starting some shit with the Mongols, they soon melted away. Incredibly, despite the outrage in Otrar, Genghis was willing to give him a second chance. The trade deal was still a priority for the Mongols (for starters, their campaign of conquest had not been great for the agriculture of their homelands, so they needed to buy stuff). And so Genghis sent three envoys—one Muslim, two Mongol—to set things straight with Muhammad, demanding punishment for Inalchuq, compensation for their goods and a return to peace.

Instead of apologizing, the shah beheaded the Muslim envoy and burned the beards off the faces of the Mongols, sending them back to Genghis mutilated and humiliated.

Why? I mean, literally, why would you do that? Did Muhammad really start a war with Genghis Khan because he thought that a description of where the sun sets was a diss?

It's certainly possible, and not significantly dumber than any other explanation. But at the same time, it's worth noting that Muhammad's paranoia extended further than a bit of extremely fragile masculinity. Of Turkic origin and descended from a slave, he was often looked down upon by neighboring Persian and Arabic nobles in the Muslim world. His empire was almost as young as Genghis's, and internally divided. He had a difficult relationship with his mother, which never really helps. He also had a long-running beef with al-Nasir, the Arab caliph of Baghdad, who he now suspected of secretly plotting with the Mongols to bring him down. (In fairness, it's not impossible that al-Nasir actually *was* plotting with the Mongols, although it would have been a pretty counterproductive move for all concerned.) And a failed attempt to capture Baghdad in 1217, when Muhammad's troops got lost in the snow as they tried to cross some mountains, had probably left him feeling even more raw about his military prowess.

In addition, he may have simply underestimated the threat Genghis posed. In a good example of why you should wait until you've got as much information as possible before doing anything rash, as the now-beardless Mongol envoys were heading home with news of Muhammad's provocation, one of the shah's emissaries was headed the other way carrying news of exactly how strong the Mongol forces were. Upon finding out what he'd just put himself up against, the shah's reaction seems to have been, to paraphrase roughly, "Oh."

And so Genghis climbed to the top of Burkhan Khaldun, the mountain near his birthplace that he always went to when contemplating war, and prayed for three days and nights. Then he sent one final message to Muhammad—and this time, at least, it was straightforward enough that it couldn't really be misinterpreted. "Prepare for war," he told the shah. "I am coming against you with a host you cannot withstand."

Genghis set out for Khwarezm with his army in 1219. By 1222, the Khwarezmian Empire had been wiped off the map.

Estimates vary wildly, but it seems likely that the Mongols had just over 100,000 troops, while the shah had twice as many or more, and was fighting on familiar terrain. Didn't matter. Muhammad threw away his home advantage by deciding to try and wait out the Mongol forces behind well-defended city walls, in the belief that they were rubbish at sieges. In fairness, they *had* been rubbish at sieges, but what Muhammad didn't appreciate was that the Mongol army were extremely quick learners. The first siege of the war (against the city of Otrar, naturally) lasted for months. After that, most of the rest lasted weeks, or days.

The Mongol army was agile, adaptable, disciplined and intelligence-driven. Genghis split his forces up to attack from unexpected directions, cut off backup or take on multiple targets at once. They prioritized speedy communication and changed their tactics easily, assimilating strategies and weaponry from those they had conquered. And they were utterly, utterly ruthless.

They swept through Khwarezm with terrifying speed. Every city they took was given a chance to surrender, and the ones that did were treated with relative generosity (emphasis on the "relative"): they'd be looted for everything they owned, sure, but most of the population would be allowed to live. But if they didn't surrender, or if they tried to rebel later, then the response was brutal.

In Omar Khayyam's birthplace of Nishapur, where Genghis's favorite son-in-law was killed in battle, his grieving widow was allowed to choose the city's fate: as a result, every single person in the city (a few skilled artisans aside) was executed, their 17,000 skulls piled into enormous pyramids. The slaughter took 10 days, after which the Mongols killed every dog and cat in the city, as well, just to really emphasize the point. In Gurganj, one of the few cities that managed to hold them off for several months, they opened the dam that held back the diverted Amu Darya river, sending down a deadly wave of water that wiped out the city entirely (and reputedly changed the course of the river for several centuries, as mentioned a few chapters back). Both of these events happened in the same month of 1221, by the way, which must make it one of the more destructive months in history.

Genghis knew the propaganda value of terror, and found that the highly literate Islamic world was a big help there: he liked to ensure that letters were sent telling tales of his conquests, as it increased the chances of the next few cities surrendering without a fight.

At the same time, he also took care to be respectful of religion, often treating particularly holy sites more gently. For all its wild brutality, the Mongol Empire under Genghis was also surprisingly tolerant, to the point that he created possibly the world's first ever law enshrining freedom of religion. This had pragmatic benefits, of course: it was easier for opponents to see the benefits of surrender if they knew they weren't fighting a holy war, and it turned religious minorities everywhere into potential allies. When the city of Bukhara, a center of Muslim theology, fell in the early months of 1220, Genghis ordered that amid the destruction, the Great Mosque be left untouched. He even visited the mosque himself—the only time in his life that he's recorded as actually entering a city he had conquered. A big fan of tents and open plains, whose own god was the Eternal Blue Sky, Genghis never really saw the point of cities, other than as things to conquer.

And what of Muhammad, whose jaw-dropping diplomatic incompetency was the catalyst for all this? Holed up in Bukhara's sister city of Samarkand, the shah saw that the writing was on the wall as soon as Bukhara fell. He fled the city, and spent the next year engaged in what could be generously described as "fighting a rearguard action," or less generously as "running away." Genghis devoted 20,000 troops to pursuing him across his crumbling empire, with orders not to return until they'd caught or killed him. They chased him as far as the shores of the Caspian Sea, where he sought refuge on a series of islands. It was on one of those islands that—by now penniless, dressed in rags and losing his mind—Muhammad died of pneumonia in January 1221.

If Genghis had stopped his attacks once the cause of his ire was dead, then maybe Muhammad's name would only be a historical footnote today. The trouble was, he didn't stop. The destruction of Khwarezm continued throughout 1221, and the violence became ever more extreme. The orders to wipe out entire populations of resisting cities became explicit, as Nishapur, Gurganj, Merv and others were to find out.

And once the Khwarezmian Empire had been obliterated, Genghis...just carried on, possibly impressed by how easy it had all been. His original lack of interest in extending his empire westward had now transformed into a very strong desire to see how much more he could conquer. Much of the Asian Islamic world was gobbled up, and the Mongols pushed on into Europe. After Genghis died in 1227, his sons and grandsons continued the expansion. At its height, the Mongol Empire was the largest land empire the world has ever seen, stretching from Poland to Korea.

While it fractured after a couple of generations, descending into factionalism and in-fighting as empires often tend to do, its legacy continued in some areas for far longer—even into the twentieth century. In the emirate of Bukhara, the direct descendants of Genghis ruled until as recently as 1920, the last reign of the Khan dynasty finally ending only when the Bolsheviks came along. (In 1838, a British soldier named Charles Stoddart, on a diplomatic mission to win Bukhara over to the cause of Britain's own empire, ironically managed to re-create Muhammad's folly in microcosm: casually insulting Emir Nasrullah Khan for no apparent reason, he was thrown into a deeply unpleasant place known as the Bug Pit, where he spent several horrifying years having his flesh eaten by insects before finally being executed. Do not mess with the Khans.)

The culture and history and writings of many of the places the Mongols conquered were completely destroyed, entire populations were displaced and the death toll runs into uncountable millions. There is an upside, sort of: the unification and stabilization of the very trade routes that kicked off the whole affair brought about a continent-spanning cultural exchange that helped jump-start the modern age across much of Eurasia. The downside to that is that they exchanged diseases as well as culture, including the bubonic plague, which killed millions more.

And all because a man with a fragile ego decided that diplomacy was for losers, and that a simple request for a trade deal had to be some kind of nefarious plot. Ala ad-Din Muhammad, you fucked up, my son.

4 MORE IMPRESSIVE FAILURES OF INTERNATIONAL RELATIONS

Atahualpa

Inca ruler who in 1532 made a similar mistake to Moctezuma when faced with a Spanish incursion, except he improved on it by getting drunk before meeting the Spanish, and leading his troops into a really obvious trap.

Vortigern

Fifth-century British ruler who—lacking defenses against the Picts following Roman withdrawal—reputedly invited Saxon mercenaries to stay in Britain to fight for him. The Saxons decided to just take over instead.

Francisco Solano López

Paraguayan leader who managed to get his relatively small country into a war with the much larger countries of Brazil, Argentina and Uruguay. It's estimated that more than half of his country's population died.

Zimmerman Telegram

In 1917, Germany sent a secret telegram to Mexico offering a military alliance if the USA joined World War I—and promising them Texas, New Mexico and Arizona. When the British intercepted it, all it did was encourage the US to join the war (and Mexico wasn't even interested).

THE SHITE HEAT OF TECHNOLOGY

The human compulsion to explore and to always seek out new horizons is—as I think we've mentioned already—one of our defining characteristics. It was that urge to explore and to uncover new knowledge that drove NASA to launch the Mars Climate Orbiter into the vast, empty black void of space in 1998.

A few months later, the Mars Climate Orbiter ended up crashing onto a load of rocks, like an idiot.

In a spectacular demonstration of humanity's ability to make essentially the same mistakes over and over again, a little more than five centuries after Christopher Columbus messed up his units of measurement, got his sums wrong and ended up running aground on the Americas, the people behind the Orbiter messed up their units of measurement, got their sums wrong and ended up falling to the ground on Mars.

Humanity's next great step on our journey through history, the scientific revolution, began in the sixteenth century in the letters and books being exchanged by philosophers across Europe. To begin with, it wasn't really so much a revolution as a catch-up session; quite a lot of it was just rediscovering knowledge that had already been worked out by previous civilizations. But hand in hand with the rise of global travel, conquest and trade—always hungry for new knowledge and new technology—over the next few centuries it produced a huge expansion in our understanding of the world. It didn't just give us lots of science, it gave us the idea of science itself, as something that was a distinct discipline with its own methods rather than just being one variant of "having a bit of a think."

The pace of technological change continued to accelerate until, in towns across northern Britain in the seventeenth and eighteenth centuries, fueled by cheap American cotton from slave plantations, another revolution started taking place. This time it was in manufacturing methods, with the rise of the machines enabling production on a mass scale—something that would spread around the world and change forever our cities, our environment, our economies and our ability to order a foot spa off Amazon at 3:00 a.m. while drunk.

The dawn of the scientific, technological and industrial ages have brought opportunities to humanity that our ancestors could never have dreamed of. They have also, unfortunately, offered us the chance to fail on a scale never previously anticipated. When Columbus got his units of measurement wrong, he was at least confined to making his mistakes on the surface of the earth. Now, as the unfortunate story of the Mars Climate Orbiter shows, we get to screw up in *space*.

The failure of the Orbiter only started to become apparent

several months into the mission, when attempts by mission control to make minute adjustments to the spacecraft's trajectory in order to keep it on course started to not quite have the effect they were intended to. But quite how wrong it had gone only became apparent when the craft reached Mars and attempted to go into orbit, only to lose contact with ground control almost immediately.

The investigation afterward revealed what had happened: the Orbiter was using the standard metric unit of Newton seconds to measure impulse (the total amount of thrust applied in a maneuver). But the software on the ground computer, supplied by a contractor, was using imperial measurements of pound seconds. Every time they'd fired the ship's engines, the effect had been more than four times as much as they'd thought—with the result that the Mars Orbiter ended up over a hundred miles closer to the surface of Mars than it was supposed to. As it tried to go into orbit, it instead hit the atmosphere hard, and the cutting-edge $327 million spacecraft broke into pieces almost instantly.

That must have been embarrassing for NASA, but maybe they were able to take comfort from the fact that they're hardly alone in the field of scientific and technological cock-ups. Another example comes not from the space race, but an entirely different kind of race that scientists across the USA found themselves in when, in 1969, they were competing with their Soviet counterparts to uncover the mysteries of a revolutionary discovery: an entirely new form of water.

It was the height of the Cold War, and that all-consuming ideological showdown wasn't just playing out in geopolitical maneuvers, nuclear brinksmanship and the shadowy world of espionage. It also birthed a contest between the communist and capitalist worlds to demonstrate their scientific and engineering prowess. New discoveries and technological breakthroughs were coming at a dizzying rate, and there was a constant terror of falling too far behind the enemy; in July that year, a human would walk on the surface of the moon, put there by the American government's shocked reaction to a series of Soviet spacefaring firsts.

Amid all these grand, cinematic breakthroughs, a novel form of water initially appeared to be little more than a minor wrinkle. First discovered in 1961 by Nikolai Fedyakin, a scientist working at a provincial Soviet laboratory far from the major centers of science, it wasn't until his work was spotted by Boris Deryagin of the Institute for Physical Chemistry in Moscow that its potential importance was realized. Deryagin quickly replicated Fedyakin's work and, unsurprisingly, gladly started taking credit for the discovery—but still, outside the Soviet Union, there was little interest. It was only when he presented his findings at a conference in England in 1966 that the international community sat up and began to take note. The race was on.

At first referred to as either "anomalous water" or "offspring water," the discovery had remarkable properties. Fedyakin and Deryagin found that the process of condensing or forcing regular water through supernarrow, ultrapure quartz capillary tubes had somehow caused it to rearrange itself, radically altering its chemical properties. Anomalous water didn't freeze at 0°C; instead it froze at –40°C. Its boiling point was even more extreme, at least 150°C or possibly more, maybe as high as 650°C. It was more viscous than water, barely a liquid at all, thicker and greasier—some descriptions said it resembled Vaseline. If you cut into it with a blade, the mark would remain.

First in England, and then in the USA, scientists set about replicating the Soviets' work. It was a difficult process, as the capillaries necessary for the process also meant that only tiny amounts could be manufactured at a time: some laboratories couldn't get the technique right at all, while others raced away, producing ever larger amounts of the anomalous water. It was from one of these labs in the US that the next big breakthrough came: enough anomalous water was synthesized that they were able to perform an infrared spectral analysis of the substance. Their results were published in the prestigious journal *Science* in June 1969, one month before Armstrong walked on the moon, and the paper sent the scramble for research into the substance into overdrive. Not only was it confirmation of the water's radically different properties compared with standard water, it provided an explanation for it: the results suggested that this was a polymer version of water, the individual H_2O molecules joining up in large chained lattices that made it more stable. And so "anomalous water" became known instead by the name we know it today: "polywater."

The discovery of polywater "is sure to revolutionize chemistry," wrote *Popular Science* in December of 1969, talking at length about its possible uses in cooling systems, as a lubricant for engines or as a moderator in nuclear reactors. It explained many aspects of the natural world: polywater was found in clay, explaining why clay retains its paste-like malleability until fired at superhigh temperatures sufficient to finally remove the polywater. Polywater might be responsible for aspects of the weather, small amounts of it seeding the formation of clouds. And it was certainly present in the human body.

The discovery would likely lead to a whole new branch of chemistry, as some labs reported that they had managed to produce polymer versions of other chemically vital liquids: polymethanol, polyacetone. Or, more sinister, there were concerns that it could have military applications, even be a weapon in its own right: its structure suggested that it existed at a lower energy state than normal water, raising the possibility that polywater coming into contact with ordinary water could potentially trigger a chain reaction, inducing the everyday water to rearrange itself, too, and adopt the polymer form. One drop of polywater added to a key strategic reservoir or river, it was theorized, had the potential to gradually convert the entire body of water, turning the whole thing into a syrup. The water supply of whole countries could be sabotaged.

In the wake of the *Science* paper, the US government stepped in. CIA agents debriefed researchers involved in its study, keen to ensure that all breakthroughs were kept in American hands. Polywater was nervously discussed across the media from the *New York Times* to small-town newspapers: Was the USA falling behind the Soviets? Polywater research was prioritized, and funding set aside. Hundreds of scientific papers were published on it in the year of 1970 alone. "Good news," a relieved *Wall Street Journal* wrote in 1969, in the wake of the initial funding, "the US has apparently closed the polywater gap, and the Pentagon is bankrolling efforts to push this country's polywater technology ahead of the Soviet Union's."

You've probably guessed by now, right? I mean, we're quite a long way into this book; it should be fairly obvious at this point that the polywater story doesn't end with a scientific triumph, everybody patting each other on the back and Nobels all around. But it wasn't until the early 1970s, after years of research by the finest scientists in the very best laboratories across multiple continents, that the truth became apparent:

There's no such thing as polywater. It simply doesn't exist.

What Fedyakin and Deryagin had actually discovered, and what scientists across the world had spent years pursuing and faithfully replicating and studying every possible way, was a substance that's more accurately described as "dirty water." All of polywater's allegedly miraculous properties turned out to simply be impurities that had crept into the supposedly sterile equipment.

One skeptical American scientist, Denis Rousseau, managed to replicate the spectral analysis of polywater almost perfectly with a few drops of his own sweat, squeezed out of his T-shirt following a handball match. That's the mysterious substance the great powers of the Cold War era had been so desperately scrambling for control of. Sweat.

Awkward.

It's not like there hadn't been plenty of skeptical voices—numerous scientists felt the discovery sounded implausible; one even announced that if polywater turned out to be real he would quit chemistry entirely. But it's often hard to disprove something, especially when there's the lurking fear that the reason your polywater isn't doing what polywater is supposed to do is that you simply didn't make it properly in the first place. The difficulty of making more than trace amounts of polywater, added to the febrile atmosphere of Cold War–era scientific research, allowed scientists spread over several continents to simply see what they'd been told to expect, and to dramatically over-interpret vague or contradictory results. The whole affair was science by wishful thinking.

Even after the first papers pushing back at the existence of polywater were published (also in *Science*, in 1970), it was years before everybody finally admitted that the whole thing had been a mistake. Ellison Taylor, one of the skeptics who had been involved in finally disproving polywater, wrote in the Oak Ridge National Laboratory's in-house magazine in 1971: "[We] knew they were wrong from the beginning, and I suppose lots of people who never got involved knew it, too, but none of the chief protagonists has given any sign of admitting it." *Popular Science* even ran an article entitled "How You Can Grow Your Own Polywater" in June of 1973 (subtitle: "Some experts claim this rare substance doesn't exist. Yet here's how you can harvest enough of it for your own experiments").

It's far from the only time something like this has happened. Of course, the early centuries of science (even before the term *science* was invented) were full of popular theories that turned out to be completely wrong—in the eighteenth century it was phlogiston, the mysterious substance that lurked inside all combustible things and was released when they burned; in the nineteenth, luminiferous ether, an invisible substance permeating the universe through which light was transmitted. But those have the distinction of at least being attempts to explain something that couldn't be explained with the science of the time. Which, more or less, is kind of how science is supposed to work.

The reason science has a fairly decent track record is that (in theory, at least) it starts from the sensible, self-deprecating assumption that most of our guesses about how the world works will be wrong. Science tries to edge its way in the general direction of being right, but it does that through a slow process of becoming progressively a bit less wrong. The way it's supposed to work is this: you have an idea about how the world might work, and in order to see if there's a chance it might be right, you try very hard to prove yourself wrong. If you fail to prove yourself wrong, you try to prove yourself wrong again, or prove yourself wrong another way. After a while you decide to tell the world that you've failed to prove yourself wrong, at which point everybody else tries to prove you wrong, as well. If they all fail to prove you wrong, then slowly people begin to accept that you might possibly be right, or at least less wrong than the alternatives.

Of course, that's not how it *actually* works. Scientists are no less susceptible than any other humans to the perils of just assuming that their view of the world is right, and ignoring anything to the contrary. That's why all the structures of science—peer review and replication and the like—are put in place to try and stop that happening. But it's extremely far from foolproof, because groupthink and bandwagon-jumping and political pressure and ideological blinders are all things in science, as well.

That's how you can get a load of scientists at different institutions in different countries all convincing themselves they can see the same imaginary substance. The saga of polywater isn't alone there: six decades earlier, the scientific world had been

René Prosper Blondlot (1849–1930)

gripped by the discovery of a whole new type of radiation. These remarkable new rays (which it would eventually turn out were entirely imaginary) were called N-rays.

N-rays were "discovered" in France, and they took their name from the town of Nancy, where the scientist who first identified them worked—René Prosper Blondlot, an award-winning researcher who was widely acclaimed as an excellent, diligent experimental physicist. This was 1903, less than a decade after the discovery of X-rays had sent waves through the field, so people were primed to expect that new forms of radiation could be discovered here, there and everywhere. What's more, just as with polywater, there was more than a little international rivalry at play—X-rays had been discovered in Germany, so the French were eager for a piece of the action.

Blondlot first uncovered N-rays by accident—in fact, it was while he was conducting research on X-rays. His experimental equipment involved a small spark that would grow brighter when the rays passed by, and his attention was caught when he saw the spark flare up at a time when no X-rays could possibly be affecting it. He dug deeper, gathered more evidence and in spring 1903 announced his discovery to the world in the *Proceedings of the French Academy*. Fairly quickly, a large part of the science world went N-ray crazy.

Over the next few years, more than 300 papers would be published about the remarkable properties of N-rays by over 120 scientists (Blondlot himself published 26 of them). The qualities that N-rays demonstrated were certainly...intriguing. They were produced by certain types of flame, a heated sheet of iron and the sun. They were also produced by living things, Blondlot's colleague Augustin Charpentier found: by frogs and rabbits, by bicep muscles and by the human brain. N-rays could pass through metal and wood, and could be transmitted along a copper wire, but were blocked by water and rock salt. They could be stored in bricks.

Unfortunately, not everybody was having quite as much success in producing and observing N-rays. Many other reputable scientists couldn't seem to summon them into existence at all, despite Blondlot being very helpful in describing his methods. Possibly this was because they were hard to detect: by this point Blondlot had moved on from detecting them with a glowing spark, instead using a phosphorescent sheet that would glow faintly when exposed to the rays. The trouble was that the change in the sheet's glow was so faint that it was best seen in an entirely darkened room, and only then after the experimenter had allowed their eyes to adjust to the darkness for about 30 minutes. Oh, and it worked best if you didn't look at the sheet directly, but instead out of the corner of your eye.

Because of course there's no way that sitting in a dark room for half an hour then looking at a very faint glow in your peripheral vision would possibly make your eyes play tricks on you.

The N-ray skeptics, of whom there were many, couldn't help but notice one rather telling feature of the N-ray mania: virtually all the scientists who'd been able to produce the rays were French. There were a couple of exceptions in England and Ireland; nobody in Germany or the USA had managed to see them at all. This was starting to cause not just skepticism, but outright irritation: while the French Academy awarded Blondlot one of the top prizes in French science for his work, one leading German radiation specialist, Heinrich Rubens, was summoned by the kaiser and forced to waste two weeks trying to re-create Blondlot's work before giving up in humiliation.

All of this prompted one American physicist, Robert Wood, to pay a visit to Blondlot's lab in Nancy while he was visiting Europe for a conference. Blondlot was happy to welcome Wood and demonstrate his latest breakthroughs; Wood had a slightly different plan in mind. One of the strangest properties of the mystery rays was that, just as light is refracted through a glass prism, N-rays could apparently be refracted through an aluminum prism, producing a spectrum of ray patterns on the sheet. Blondlot eagerly demonstrated this to Wood, reading out to him the measurements of where the spectrum patterns fell. Wood then asked him if he'd mind repeating the experiment, and Blondlot readily agreed, whereupon Wood introduced a proper scientific control—or to put it another way, played a pretty funny trick on Blondlot.

In the darkness, without Blondlot noticing, he reached out and simply pocketed the prism. Unaware that his equipment was now missing its vital component, Blondlot continued to read out wavelength results for a spectrum that shouldn't have been there anymore.

Wood summarized his findings in a politely brutal letter to *Nature* in the autumn of 1904: "After spending three hours or more in witnessing various experiments, I am not only unable to report a single observation which appeared to indicate the existence of the rays, but left with a very firm conviction that the few experimenters who have obtained positive results have been in some way deluded." After that, interest in N-rays collapsed, although Blondlot and a few other true believers kept on plugging away, determined to prove that they hadn't just been studying a mirage all this time.

The stories of both polywater and N-rays are cautionary tales about how even scientists can fall prey to the same biases that affect us all, but they're also tales of science...well, working. While the hype around them both was, in retrospect, more than a little embarrassing for quite a lot of highly qualified professionals, neither mania lasted for more than a few years before skepticism and the need for hard evidence won out. Go, team.

But if these examples are relatively harmless, there have been plenty of instances where dodgy science has done a lot more than merely leave some people with bruised reputations. Like, for example, the legacy of Francis Galton.

Francis Galton was undoubtedly a genius and a polymath, but also a creepy weirdo who had terrible ideas that led to dreadful consequences. A half cousin of Charles Darwin, he achieved breakthroughs in multiple disciplines—he was a pioneer of scientific statistics, including inventing the concept of correlation, and his creations in fields as diverse as meteorology and forensics are still with us today, in the form of the weather map and the use of fingerprints to identify people.

He was obsessed with measuring things and applying scientific principles to just about everything he came across—his letters printed by *Nature* include one estimating the total number of brushstrokes in a painting (after he got bored in lengthy sittings for a portrait), and another in 1906 entitled "Cutting a Round Cake on Scientific Principles" (in short: don't cut wedges, cut straight slices through the middle, so you can push the remaining halves together to stop them drying out).

But this obsession went further than coming up with extremely British teatime life hacks. In one of his more infamous investigations, Galton toured the towns and cities of Britain in an attempt to create a map of where the women were most attractive. He would sit in a public space and use a device concealed in his pocket called a "pricker"—a thimble with a needle in it that could puncture holes in a piece of marked paper—to record his opinion of the sexual desirability of every woman who walked past. The end product of this was a "beauty map" of the country, much like his weather maps, which revealed that the women in London were the most attractive, while the women in Aberdeen were the least attractive. At least, according to the tastes of a pervy statistician furtively making notes on women's fuckability with a needle hidden in his pocket, which perhaps isn't the most objective of measures.

It was that same combination of qualities—a compulsion to measure human traits and a complete lack of respect for the actual humanity of the people being measured—that led Galton to his most infamous contribution to the world of science: his advocacy of, and indeed coining of the term, *eugenics*. He believed firmly that genius was entirely inherited, and that a person's success came from their inner nature alone, rather than fortune or circumstance. And so he believed that marriages between people deemed suitable for breeding should be encouraged, possibly with monetary rewards, in order to improve the stock of the human race; and that those who were undesirable, such as the feebleminded or paupers, should be strongly discouraged from breeding.

In the early part of the twentieth century, there was worldwide uptake of the eugenics movement, with Galton (now near the end of his life) seen as its hero. Thirty-one US states passed compulsory sterilization laws—by the time the last had finally been repealed in the sixties, over 60,000 people in mental institutions in the United States had been forcibly sterilized, the majority of them women. A similar number were sterilized in Sweden's efforts to promote "ethnic hygiene," where the law wasn't repealed until 1976. And of course in Nazi Germany... well, you know what happened. Galton would no doubt have been horrified if he'd lived long enough to see what was being done in the name of the "science" he created, but that doesn't make his original ideas any less wrong.

Or there's Trofim Lysenko, the Soviet agricultural scientist whose profoundly bad ideas contributed to famines in both the USSR and (as mentioned way back in Chapter 3) China. Unlike Galton, Lysenko doesn't even have actual legitimate scientific advances to even out his legacy. He was just inordinately wrong.

Lysenko came from a poor family, but quickly rose through the ranks of Soviet agronomy thanks to some early successes in stimulating seeds to grow without needing to be planted over the cold winters. He eventually became a favorite of Stalin, which gave him enough power to start imposing his ideas on the rest of the Soviet scientific community.

Those ideas weren't right—they weren't even close to being right—but they did have the advantage of appealing to the ideological biases of Lysenko's communist overlords. Despite the fact that genetics was a pretty well-established discipline by the 1930s, Lysenko rejected it entirely, even denying that genes existed, on the grounds that this promoted an individualistic view of the world. Genetics suggested that organisms' behaviors were fixed and unchanging, while Lysenko believed that changing the environment could improve the organism and pass those improvements down to its offspring. One species of crop could even turn into another, given the right environment. Rows of crops should be planted more closely together, he instructed farmers, because plants of the same "class" would never compete with each other for resources.

None of these things was true, and what's more they were very obviously not true, as evidenced by the fact that attempts to put them into practice just ended up with a lot of dead crops. That didn't stop Lysenko maintaining his political power and shutting down any criticism—to the point of having thousands of other Soviet biologists sacked, imprisoned or even killed if they refused to abandon genetics and embrace Lysenkoism. It wasn't until Khrushchev was forced out of power in 1964 that other scientists finally managed to persuade the party that Lysenko was a charlatan, and he was quietly shuffled out. His legacy was to contribute to millions of deaths, and to set the field of biology in the Soviet world back decades.

But if Lysenko's mistakes in biology were entirely enabled by communism, the next case was pure capitalism—the tale of a man who managed to make not one, but two of the most disastrous mistakes in the history of science, all within the space of one decade.

Lead Astray

In 1944, the genius engineer, chemist and inventor Thomas Midgley Jr., a man whose discoveries had helped shape the modern world to a remarkable degree, died at home in bed at the age of 55.

Dying at home in bed sounds quite peaceful, you'd think. Not in this case. Paralyzed below the waist due to a bout of polio some years earlier, Midgley disliked the indignity of being lifted in and out of bed, and had put his talent for innovation to good use, building himself an elaborate system of pulleys so he could do it himself. Which was all going terribly well until that day in November, when something went a bit wrong and he was found strangled to death by the ropes of his own device.

Thomas Midgley Jr. (1889–1944)

The manner of his death is grimly ironic enough—but that's not the reason Tom Midgley is in this book. He's in this book because, incredibly, being killed in bed by his own invention doesn't even make it into the top two biggest mistakes of his life.

In fact, by pretty much any standard, he has to rank as one of the most catastrophic individuals who ever lived.

Midgley was a quiet, clever man who spent most of his life in Columbus, Ohio. From a family of inventors, he had barely any training as a chemist, but showed a knack for problem-solving across a range of disciplines—through a combination of systematic examination of the issues on one hand, and on the other a tendency to haphazardly but doggedly throw solutions at a problem until something stuck.

In the 1910s and 1920s, he was working on the problem of car engines "knocking"—a persistent problem where engines would stutter and jerk, especially when put under strain. This didn't just make early automobiles kind of suck, it also reduced fuel efficiency, a major concern at a time when there were early worries that the world's oil supplies were due to run low sooner rather than later.

Midgley and his boss, Charles Kettering, suspected that knocking was down to the fuel used burning unevenly, rather than a fundamental flaw in the design of engines. So they set about trying to find an additive that would reduce this effect. Initially, for reasons that make astonishingly little sense, they settled on the idea that the solution was "the color red." Midgley went out to get some red dye, but the lab didn't have any. He was told, however, that iodine was kind of reddish and dissolved well in oil, so he basically went, "Ah, what the heck," stuck a load of iodine in some gasoline and whacked it into an engine.

It worked.

It was complete dumb luck, but they'd hit on proof that they were on the right track. Iodine itself wasn't a workable solution: it was too expensive and too difficult to produce in the quantities they'd need. But it was enough to convince them to carry on their work. Over the following years, they tried—depending on which corporate statement you believe—somewhere between 144 and 33,000 different compounds. If that seems like quite an imprecise range, well, there's a reason why the companies behind their work have been kinda vague about the research process.

The reason is that the substance they finally settled on was lead (specifically, a liquid compound called tetraethyl lead, or TEL). And lead is a deadly poison. It causes, among other things, high blood pressure, kidney problems, fetal abnormalities and brain damage. It particularly affects children.

Midgley's story is often told as an example of "unintended consequences," which... No, not really. Granted, "poisoning entire generations of people all across the globe" wasn't actually his goal. But equally, nobody involved in the production and popularization of leaded gasoline gets to play the "oh no, what a horrible and unforeseen surprise" card.

Lead's toxic nature wasn't a new discovery—it's been known

for literally thousands of years. Before the first gas pump had even started supplying the new antiknock fuel in early 1923, medical experts were warning that this was a terrible, terrible idea. William Clark of the US Public Health Service wrote in a letter that using tetraethyl lead presented a "serious menace to public health" and predicted—entirely accurately—that "on busy thoroughfares it is highly probable that the lead oxide dust will remain in the lower stratum."

In an even more upsettingly accurate prediction in 1924, a leading toxicologist foresaw that "the development of lead poisoning will come on so insidiously that leaded gasoline will be in nearly universal use...before the public and the government awaken to the situation."

And the thing is, it's not like lead was the only available solution. In the years since their iodine breakthrough, Midgley's team had come up with *loads* of effective antiknock agents. One of which was impressive in its simplicity: ethanol. A viable fuel in its own right, your basic drinkin' alcohol isn't just good for sterilizing physical wounds and temporarily cleansing emotional wounds, it also works well as an antiknock additive—with the added benefit that it's incredibly easy and cheap to produce on a mass scale.

In fact, for years, Midgley's team had been backing ethanol as the perfect solution to the engine-knocking problem. So why did they drop that in favor of a substance that everybody knew was toxic as hell? You will be shocked to learn that the reason was money.

The trouble was that ethanol was simply *too* easy and cheap to produce. And, crucially, it wasn't patentable. Charles Kettering's company, Delco, had been bought up by the giant General Motors in 1918, and there was pressure on his research team to show that they could generate actual cash, rather than just a bunch of pie-in-the-sky tinkering. Ethanol—a substance so easy to make that people could do it at home, with no hope of it being turned into a proprietary product—was useless for that purpose. And so they went with lead.

In case you think poor Thomas Midgley was simply a harmless inventor whose work was misused by some nasty plutocrats: nah. In fact, he was the one who suggested and strongly advocated using lead. He even did the math, calculating that they could charge an extra three cents on the gallon on TEL fuel, and predicting they could capture 20 percent of the gasoline market with an aggressive ad campaign. On that, like a lot of things, he was wrong by virtue of wildly underestimating his work's impact: in just over a decade, tetraethyl lead gasoline—under the brand name Ethyl, cunningly not mentioning the "lead" bit—had actually captured 80 percent of the US market.

All the way, General Motors and Midgley insisted it was safe, despite plenty of what you might call "warning signs." Huge flashing neon warning signs. Like the fact that in February 1923, when Ethyl first went on sale, Midgley himself had to take the whole month off work due to ill health caused by the lead fumes. Or like the fact that workers at the factories that made the fuel kept on dying a lot. Five workers died from lead poisoning at the Bayway plant in New Jersey, and 35 were hospitalized, many of them driven insane by the neurological effects of lead—"the patient becomes violently maniacal, shouting, leaping from the bed, smashing furniture and acting as if in delirium tremens," one report recorded. Six workers died at the New Jersey Deepwater plant, where hallucinations caused by the lead were so common that workers renamed it the "House of Butterflies." The deaths made it to the front page of the *New York Times*. Faced with a public relations crisis, sale of Ethyl was suspended, and the US Surgeon General hastily set up a committee to determine its safety.

And then, in a remarkable bit of corporate judo that would come to serve as the template for a whole range of fucking-stuff-up industries over the rest of the twentieth century, the companies behind the Ethyl Gasoline Corporation—General Motors, Standard Oil and chemical behemoth DuPont—managed to turn that PR crisis into a PR win.

It was a classic example of responsible people answering entirely the wrong question. The focus of public concern on the deaths at the manufacturing stage was so strong that, in the end, that was the only issue the Surgeon General's committee actually delivered a verdict on. Persuaded by the firms' assurances that extra safety measures would be taken in their factories—TEL, Midgley said in his testimony, was "not so much a dangerous poison as it is a treacherous one"—the committee decided not to ban its manufacture. The much bigger question of its effect on the exhaust-fume-breathing public was never actually decided: that was, in the time-honored tradition, a matter for future research. But the committee's decision was spun to the public and the politicians as giving leaded gasoline an entirely clean bill of health.

In case you're wondering about that "future research," for the next four decades, almost all of it was either funded by the companies who made leaded gasoline, or carried out by their own staff. In shock news, this research was inconclusive! Which was all the producers of TEL needed to make the case that the question was still undecided, and it would be very, very bad and wrong to stop selling this lovely fuel that had enabled so many dreams to come true.

Because once leaded gasoline had been given the supposed all clear, the sky was the limit. Not only did it stop car engines knocking, it enabled the development of a whole new generation of more powerful engines, which turned cars from practical but ungainly old bangers into fast, smooth, sleek objects of desire. An aggressive advertising campaign played on fears of having a slow, crappy car if you didn't use leaded fuel; rival products from competitors, including those that used ethanol—the very substance Midgley's team had advocated for years—

were derided as substandard. When health fears were raised in other countries about the introduction of leaded fuel there, the fact that the Americans had said it was A-OK was used to tamp down those fears; the Surgeon General, Hugh Cumming, even communicated with his counterparts overseas to tell them how extremely safe it was.

Backed by some woefully bad science, a rapacious desire to make money and the fact that powerful cars are cool and let you travel farther, leaded fuel quickly became the standard around the world. Thanks to advances in the oil extraction game, the supposed fuel shortages that had prompted the work on anti-knock agents in the first place never materialized, so instead all the benefits from lead went toward making ever more powerful engines. The age of the automobile was here, and across the globe, more and more people started breathing in lead fumes.

The thing about lead is that it doesn't break down. While some toxins will become less dangerous with time, lead builds up—in the air, in the soil and in the bodies of plants and animals and humans. In 1983, a report by the UK Royal Commission on Environmental Pollution concluded that "it is doubtful whether any part of the earth's surface or any form of life remains uncontaminated by anthropogenic lead." Children's bodies are especially at risk, as they absorb five times the amount of lead into their systems as adults. In the US alone, it's estimated that 70 million children had toxic levels of lead in their blood in the decades between the 1920s and 1970s.

The effects of lead are severe. The World Health Organization estimates that hundreds of thousands of people die annually worldwide from lead-poisoning illnesses, such as heart disease. Beyond the physical health effects, lead also damages children's neurological development—it causes a drop in IQ levels among affected populations, and is estimated to be the cause of over 12 percent of developmental intellectual disabilities around the world.

It also causes behavioral problems, such as antisocial behavior, which gives rise to one of the more nightmarish possible consequences of Thomas Midgley Jr.'s work. It's important to point out that this is, to date, just an unproven hypothesis, but a number of researchers have pointed out that the enormous spike in crime levels that occurred across much of the world in the postwar period tracks pretty neatly the growth in lead pollution.

The crime levels that gave rise to many of our casual cultural assumptions—the feral teens and the inner-city hellscapes and all that nineties talk of "superpredators"—are in fact a historical anomaly, a global blip that is hard to explain and now seems (hopefully) in the past. But in country after country, regardless of their social conditions or political direction, crime started to spike a couple of decades after leaded gasoline was introduced there—in other words, when the first children to be exposed to it in significant amounts reached their teens and early twenties. And the correlation applies in the opposite direction, too: the past few decades have seen a consistent fall in violent crime around much of the world, again regardless of what social policies each country might be implementing. But the drop in crime does seem to occur around two decades after each particular area banned lead in gasoline—happening sooner in the places that banned lead earlier, and more rapidly in the places that stopped its use abruptly rather than gradually phasing it out.

To reiterate, correlation is definitely not causation, and this is still no more than informed speculation. Given the ethical issues you'd hit if you tried to inject a load of children with lead and then wait around to see how many crimes they committed twenty years later, it may never be proven one way or the other. But in addition to possibly millions dead, and the fact that we've polluted every corner of the planet, and the knowledge that multiple generations of children had poison in their blood that affected their intelligence (those are the generations, by the way, who have BEEN IN CHARGE OF THE WORLD FOR

THE LAST FORTY YEARS), the possibility that we might have caused a global crime wave that lasted for decades and entirely reshaped our view of society, simply because Thomas Midgley wanted to make three cents on the gallon, is...well, it's a very long and very dark joke.

Midgley himself did not hang around after inventing leaded gasoline. Ever the tinkerer, he quickly moved on to other areas of investigation—and he still had his second catastrophic mistake to make.

Unlike the years-long search for a better fuel, this one came quickly. In fact, according to corporate legend, it took Midgley all of three days after being set the problem before he found a solution. And unlike lead, this one genuinely is a case of unintended consequences: there were no dire warnings ignored or risks covered up. It was simply a product of assuming, in the absence of any evidence, that everything would be fine.

This time, the problem Midgley confronted was cooling things down. This was 1928, not long after the beginning of the era of mechanical refrigeration (before that, the ice-harvesting industry was big business, with vast quantities of ice being carved up and shipped from the cold parts of the world so that people in the warmer parts could keep stuff cool). The trouble was, all the substances currently being used for refrigeration were (a) expensive, and (b) extremely dangerous. They had a tendency to catch fire, or poison people in large numbers if they leaked—the year after Midgley began his work on refrigeration, a leak of methyl chloride from a Cleveland hospital's refrigeration unit killed over 100 people.

Unsurprisingly, this was causing something of a drag on widespread adoption of refrigeration technology.

The goal was simple: to find a cheap, nonflammable, nontoxic substance that would do the same job as the current refrigerants. General Motors had recently bought a refrigeration company, which they renamed Frigidaire, and they knew that if they could crack the problem, they'd make a mint.

Midgley's approach was less haphazard this time around (he'd now had over a decade's experience in chemistry, after all). Studying the chemical properties of known refrigerants, he quickly identified fluorine as a likely candidate, ideally in a compound with carbon to neutralize its toxic effects. And he pretty much nailed it out of the gate, as one of the initial substances his team created to test was dichlorodifluoromethane. These days, it's better known by the brand name they gave it: Freon.

Midgley demonstrated its safety to great acclaim at a meeting of the American Chemical Society, theatrically inhaling a lungful of it and using it to blow out a candle. Nontoxic, nonflammable and an excellent refrigerant. Perfect. Indeed, he hadn't just discovered a new compound, he'd discovered a whole new class of them, all of which had similar properties. They became known as chlorofluorocarbons—or, to use the common abbreviation, CFCs.

Unfortunately, in the early 1930s, nobody even really knew what the "ozone layer" was, or just how important that thin band of oxygen molecules in the stratosphere was in shielding the surface of the earth from the sun's harmful ultraviolet rays. They certainly didn't know that CFCs, entirely harmless at sea level, would become far more dangerous when they got into the upper atmosphere, where that same ultraviolet radiation would cause them to break down into their constituent elements—and that one of those elements, chlorine, would destroy the ozone, robbing the planet of its protective shield.

In fairness, they also didn't anticipate that the use of CFCs would end up being far wider than refrigeration. Very quickly people worked out that these new, exciting and extremely safe chemicals had lots of other uses—most notably as a propellant in aerosol sprays. In a darkly amusing bit of historical irony, during and after World War II, CFCs were widely used to spray insecticides, including that other classic example of large-scale chemistry screw-ups, the birth-defect-causing nightmare that was DDT.

After the war, aerosols really took off, in everything from spray paint to deodorants. And they took off in another, more literal way: the vast amounts we started releasing made their way upward into the stratosphere, where they set about dismantling the ozone layer.

The good news here is that this time humanity realized the problem before it could cause death on a massive scale. Woo-hoo! Score one for humans! In the 1970s (just as the first moves to phase out leaded gasoline were beginning), the growing hole in the ozone layer was also discovered, along with the link to CFCs. With that came the warning: if ozone depletion carried on at the current rate, humans would be exposed to more and more damaging UV radiation, and within a matter of decades cancer cases and blindness would soar.

And so from the 1970s to the 1990s, the world set about unwinding Thomas Midgley's legacy, as both of his major inventions were either banned or phased out in most countries around the world. We're still stuck with huge amounts of environmental lead—it doesn't simply break down or vanish, and cleaning it up is a nightmare. But in good news, at least in most places children aren't breathing it in as much anymore, and the amount of lead in the blood of many kids is now below toxic levels. Hurrah. The ozone layer, meanwhile, is slowly repairing itself now that CFCs have been widely banned: if all goes well, it should be back to pre-Midgley levels by, ooh, 2050-ish. *Go, team.*

Midgley's reputation, meanwhile, is set: he was a "one-man environmental disaster," as *New Scientist* described him; a man who in the words of historian J. R. McNeill (in his book *Something New Under the Sun*) "had more impact on the atmosphere than any other single organism in earth's history."

But it's also true to say that he shaped the modern world, often in unexpected ways. Antiknock fuel led to cars becoming the dominant mode of transport in many parts of the world, and established them not just as tools but as status objects that became a potent symbol of personal identity and individualism. CFCs didn't just bring your domestic fridge into existence, they powered air-conditioning, too, without which many major world cities simply wouldn't exist in the same form as they currently do. His two inventions even teamed up: more powerful vehicles with inbuilt AC made regular long-distance driving a realistic, even enjoyable proposition. Large swathes of the American West and much of the Middle East, to take just two examples, would likely be very different places without Thomas Midgley's creations.

There was a knock-on effect on wider culture, too—for example, in America, movie theaters were among the earliest adopters of air-conditioning, helping to boost the popularity of cinema as a leisure activity during the Great Depression, cementing the cultural impact of the golden age of film and making it perhaps the defining entertainment form of the twentieth century. Basically, what we're saying is Thomas Midgley invented Los Angeles: a city that runs on cars and AC, and is home to the movie business.

So the next time you sit in a cinema watching a dumb Hollywood film about a cop who doesn't play by the rules taking on a crime wave, just remember that almost everything about your experience may well come down to the fact that Thomas Midgley Jr. assumed the chemicals he'd invented would be harmless, and would make him three cents on the gallon.

6 SCIENTISTS WHO WERE KILLED BY THEIR OWN SCIENCE

Jesse William Lazear

American medic Jesse William Lazear proved beyond doubt that yellow fever was transmitted by mosquitoes—by letting one of the disease-carrying mosquitoes bite him. He died, proving his theory right.

Franz Reichelt

An Austrian French tailor who in 1912 confidently attempted to test his elaborate new parachute suit by jumping from the Eiffel Tower while wearing it (he was supposed to use a dummy). Plummeted to his death.

Daniel Alcides Carrión García

Peruvian medical student Carrión was determined to investigate Carrion's disease. Of course, it wasn't called Carrion's disease then. It got that name after he injected himself with blood drawn from the warts of a victim, and died.

Edwin Katskee

A doctor who in 1936 wanted to know why cocaine—then used as an anesthetic—had negative side effects. Injected himself with a ton of it, spent the night scrawling notes on the walls of his office in increasingly illegible handwriting, then died.

Carl Wilhelm Scheele

A genius Swedish chemist who discovered many elements—including oxygen, barium and chlorine—but had a habit of *tasting* each of his new discoveries. Died in 1786 of exposure to substances including lead, hydrofluoric acid and arsenic.

Clement Vallandigham

A lawyer who pioneered an early kind of forensic science. Defending an accused murderer, he proved that the supposed victim could have accidentally shot himself...by accidentally shooting himself. He died, but his client was found not guilty.

10

A BRIEF HISTORY OF NOT SEEING THINGS COMING

The modern world is, let's be honest, a confusing place.

We live at a time when technological and societal changes happen with dizzying speed. Dramatic shifts in the way we live can happen inside the space of a generation, or a decade, sometimes in less than a year. Everything seems to be constantly new: and yet, at the same time, it's hard to escape the feeling that we're just replaying the mistakes of our past at an ever-increasing rate. Somehow we consistently fail to see them coming.

As we said all the way back in the first chapter, our ability to accurately predict the future and plan for it has never been great, but the accelerating pace of change over the past few centuries hasn't exactly helped. When we're surrounded by shiny and unexpected new things all the time, those heuristics we use to make judgments get thrown out of whack. When we're bombarded by ever more information, it's not surprising if it gets too much to process and we fall back on picking out the bits that confirm our biases. How can any of us tell if we're falling victim to the Dunning–Kruger effect if we're constantly having to learn how to do new things?

And so we live in an age of endless firsts, most of which we either didn't see coming, or we ignored the people who did. And unfortunately, not all of those firsts are good. Just ask Mary Ward.

Mary Ward was a pioneer in many ways. She was born into an aristocratic family in the Irish county of Offaly in 1827, but not just any family: from a young age she was surrounded by scientists, both relatives and the visitors they received. She was lucky enough that not only did they nurture her interest in science, they also were able to fund it. As a child, seeing her interest in the natural world, her parents bought her a microscope—the best in the country at the time. It was an inspired present, because it turned out that Mary had a rare skill for drawing the specimens she observed with the microscope. (As a teenager, she also sketched the construction of the Leviathan of Parsonstown—a huge 72-inch reflecting telescope built by her cousin, former Royal Society president William Parsons—which would hold the record for the largest telescope in the world until 1917.)

As she grew into adulthood, Mary corresponded with many scientists, and her talent for illustration saw her commissioned to produce illustrations for several of their books. Then in 1857, disappointed with the quality of microscopy books on offer, she decided to print a book of her own drawings. Afraid (not without reason) that no publisher would touch it because she was a woman, she self-published 250 copies. They sold out, and the book came to the attention of a publisher, who believed that the beauty of her illustrations and the quality of her writing meant that, in this one case, perhaps the issue of her sex could be overlooked. Published as *A World of Wonders as Revealed by the*

Title page from A World of Wonders as Revealed by the Microscope *by Mary Ward, 1859*

Microscope, it became a bit of a publishing sensation—reprinted eight times over the coming decade, making it one of the first books in the category that today we might call "popular science."

That wasn't the end of her popular science career—she wrote two further books, including a telescope companion to the microscope book, which were displayed at the 1862 Crystal Palace exhibition; she would illustrate numerous other scientific works for eminent scientists; she published articles in several journals, including a well-received study of natterjack toads; and she became one of only three women permitted to be on the Royal Astronomical Society's mailing list—one of the other two was Queen Victoria. She never got a degree, though, because women weren't allowed to.

Except...all of this is preamble, because while Mary Ward was a talented woman who led a remarkable life, that's not why we remember her today. Maybe it should be. But it isn't, because of what happened in Parsonstown on August 31, 1869. On that day, at the age of 42, she and her husband, Captain Henry Ward, were riding in a steam-powered automobile. This vehicle was a homemade one—she was always surrounded by scientists, so of course it was—built by the sons of her cousin William Parsons.

Riding in a vehicle like this was a new experience at the time, an early sign of the age to come. The steam-powered automobile had been invented a century earlier in France, but this was still years before the advent of anything we'd recognize as a car today. What vehicles there were—hulking, ungainly things that were widely suspected of damaging roads—had caused enough of a sensation that Britain had passed a law regulating their use a few years earlier in 1865, but they were still rare, experimental novelties. Of the billions and billions of humans who've ever lived on this planet, Mary Ward was among the first fraction of a fraction of a fraction of a percent to ride in a car.

Records tell that as the vehicle trundled down the Mall in Parsonstown at a speed of three and a half miles an hour, it turned sharply into the corner of Cumberland Street, by the church. Maybe it was simple bad luck. Maybe the road was uneven, not being designed for anything more than a horse and cart. Maybe they weren't thinking about the concept of "turning too sharply," because cars and horses handle very differently and the risks aren't the same. Maybe Mary was simply thrilled by the experience, excited about the possibilities of the future, and she leaned out a bit too far to see the road pass beneath.

Whatever the reason, as the vehicle took the corner, one side of it tipped up slightly, and Mary was thrown from the car and under the wheels. Her neck was snapped, and she died almost instantly.

Mary Ward was the first person in the history of the world to die in a car accident.

She was a pioneer in many ways, but you don't always get to choose what you're a pioneer of. Today, around the world an estimated 1.3 million people die in car accidents every year. The future keeps on inconveniently arriving faster than we were expecting, and we keep struggling to predict it.

For example, in 1825, the *Quarterly Review* predicted that trains had no future. "What can be more palpably absurd than the prospect held out of locomotives traveling twice as fast as stagecoaches?" it asked.

A few years later, in 1830, William Huskisson, a British member of parliament and former minister of state, was attending the opening of the Liverpool and Manchester Railway. He was riding from Liverpool to Manchester in a train with the Duke of Wellington and numerous other dignitaries. Stopping off halfway for the engine to take on water, the passengers were instructed not to leave the carriages, but they did, anyway. Huskisson decided he should go and shake the Duke of Wellington's hand, as they'd had a falling-out, which is why he was standing on the opposite line when George Stephenson's famous Rocket was speeding past the other way. The passengers were warned to get out of the path of the oncoming train, but Huskisson, unfamiliar with the novel situation, panicked and couldn't decide where to move. In the end, rather than simply standing with the other passengers on the far side of the line, he instead tried clambering up onto the carriage of Wellington's train, only for the door he was desperately holding onto to swing open, putting him right in the path of the Rocket. And so William Huskisson was one of the first people in history to be killed by a train.

In 1871, Alfred Nobel said of his invention of dynamite: "Perhaps my factories will put an end to war sooner than your congresses: on the day that two army corps can mutually annihilate each other in a second, all civilized nations will surely recoil with horror and disband their troops."

In 1873, stock markets around the world crashed as a bubble of speculation finally burst. The global economic depression lasted for years.

In 1876, William Orton, the president of Western Union, advised a friend against investing in Alexander Graham Bell's new invention—the telephone—by telling him: "There is nothing in this patent whatever, nor is there anything in the scheme itself, except as a toy."

A few years after Nobel, in 1877, Richard Gatling, the inventor of the Gatling gun, wrote to a friend that he had hoped its invention would usher in a new, humanitarian era of warfare. He wrote how he was moved to invent it after he "witnessed almost daily the departure of troops to the front and the return of the wounded, sick, and dead… It occurred to me if I could invent a machine—a gun—which could, by its rapidity of fire, enable one man to do as much battle duty as a hundred, that it would, to a great extent, supersede the necessity of large armies, and consequently, exposure to battle and disease be greatly diminished."

In 1888, a Methodist missionary group in Chicago needed money and hit upon the idea of what they described as a "peripatetic contribution box"—they sent out 1,500 copies of a letter begging the recipients to send them a dime, and to forward a copy of the letter to three friends with the same request. They made over $6,000, though many people got very angry after receiving the letter multiple times. The chain letter had been born.

In 1897, the eminent British scientist Lord Kelvin predicted that "radio has no future." Also in 1897, the *New York Times* praised Hiram Maxim's invention of the fully automatic machine gun as being one so fearsome that it would stop wars from occurring, calling Maxim guns "peace-producing and peace-retaining terrors" that "because of their devastating effects, have made nations and rulers give greater thought to the outcome of war before entering upon projects of conquest."

In 1902, Kelvin predicted in an interview that transatlantic flight was an impossibility, and that "no balloon and no airplane will ever be practically successful." The Wright brothers flew their first flight 18 months later. As Orville Wright recalled in a letter from 1917: "When my brother and I built and flew the first man-carrying flying machine, we thought we were introducing into the world an invention which would make further wars practically impossible. That we were not alone in this thought is evidenced by the fact that the French Peace Society presented us with medals on account of our invention."

In 1908, Lieutenant Thomas Selfridge was a passenger in a demonstration flight piloted by Orville Wright. On the fifth circuit flying around Fort Myer in Virginia, the propeller broke and the plane crashed, killing Selfridge (Wright survived). He became the first person in history to be killed in a plane crash.

In 1912, Guglielmo Marconi, the inventor of radio, predicted that "the coming of the wireless era will make war impossible, because it will make war ridiculous." In 1914, the world went to war.

On October 16, 1929, the eminent Yale economist Irving Fisher predicted that "stock prices have reached what looks like a permanently high plateau." Eight days later, stock markets around the world crashed, as a bubble of speculation fueled by easily available debt finally burst. The global economic depression lasted for years; in the wake of the financial crisis, voters in many democracies increasingly turned to populist authoritarian politicians.

In 1932, Albert Einstein predicted that "there is not the slightest indication that [nuclear energy] will ever be obtainable."

In 1938, the British prime minister Neville Chamberlain returned home with a deal he had just signed with Adolf Hitler and predicted, "I believe it is peace for our time," before adding, "Go home and get a nice quiet sleep." In 1939, the world went to war.

In 1945, Robert Oppenheimer, the man who led the efforts to produce the atomic bomb at Los Alamos, wrote, "If this weapon does not persuade men of the need to put an end to war, nothing that comes out of a laboratory ever will." Contrary to his hopes—and the hopes of Nobel, Gatling, Maxim and Wright—we still have wars, although at least we haven't actually had a nuclear war yet (statement correct at time of writing), so Oppenheimer maybe wins this one on points.

In 1966, the eminent designer Richard Buckminster Fuller predicted that by the year 2000, "amid general plenty, politics will simply fade away."

Neville Chamberlain waves the Munich Agreement bearing his and Hitler's signatures in September 1938

A nuclear test explosion in Nevada, 1951

In 1971, the Russian cosmonauts Georgiy Dobrovolski, Viktor Patsayev and Vladislav Volkov became the first people to die in space, after their Soyuz module decompressed on their return from a space station.

In 1977, Ken Olsen, the president of the Digital Equipment Corporation, predicted that the computer business would always be niche, saying, "There is no reason for any individual to have a computer in his home." In 1978, Gary Thuerk, a marketing manager at the Digital Equipment Corporation, sent an unsolicited email plugging his company's products to around 400 recipients over Arpanet, one of the earliest manifestations of the internet. He had just sent the world's first spam email. (And according to him, it worked: DEC sold millions of dollars' worth of machines from their email campaign.)

In 1979, Robert Williams, a worker at a Ford plant in Michigan, became the first person in history to be killed by a robot.

In December of 2007, financial commentator Larry Kudlow wrote in the *National Review*: "There's no recession coming. The pessimists were wrong. It's not going to happen... The Bush boom is alive and well. It's finishing up its sixth consecutive year with more to come. Yes, it's still the greatest story never told." In December of 2007, the US economy entered recession. (At the time of writing, Larry Kudlow is currently serving as the director of the National Economic Council of the United States.) In 2008, stock markets around the world crashed as a bubble of speculation fueled by easily available debt finally burst. The global economic recession lasted for years; in the wake of the financial crisis, voters in many democracies increasingly turned to populist authoritarian politicians.

In August 2016, a 12-year-old boy died, and at least 20 other people from a nomadic group of reindeer herders were hospitalized, after an anthrax outbreak in Siberia's Yamal Peninsula. Anthrax hadn't been seen in the region in 75 years; the outbreak happened during a summer heat wave in which temperatures were 25°C above normal. The heat wave melted the thick permafrost that coats Siberia, uncovering and defrosting layers of ice that had formed decades earlier—and which held the frozen carcasses of reindeer that had died in the last anthrax outbreak in 1941.

Ice can keep pathogens preserved—alive, but in stasis—for decades, centuries, perhaps longer. The disease had been lying dormant in the subzero temperatures since the days when the Russian winter was breaking Hitler's army, just waiting for the time when its frozen cage would melt. That finally happened in

2016 (at the time, the hottest year globally since records began) as the warming world released the bacteria once more, infecting more than 2,000 reindeer before it spilled over into humans.

It's tempting to say that nobody could have foreseen a disaster so baroque, but in fact five years earlier two scientists had predicted that exactly this would happen as climate change grew worse: that the permafrost would gradually retreat, and would release long-absent historical diseases back into the world as it went. This will only continue as the temperature rises, with the curious effect of rewinding history—back past Thomas Midgley hard at work in his laboratory, back past Eugene Schieffelin standing in a park opening cages, back past William Paterson dreaming of an empire—as the cumulative effects of the Industrial Revolution unspool around us. We don't know how many people climate change will kill over the coming century, we don't know in what ways it will change our society, but we do know that at least one of its victims died because an unintended consequence of our decisions as a species was to summon zombie anthrax back from the grave. He probably won't be the last.

On May 7, 2016—a little under a century and a half after Mary Ward went for a drive one fateful summer's morning—a man named Joshua Brown was driving down a road near Williston, Florida, in his Tesla Model S, which was in autopilot mode. A later investigation showed that in the 37 minutes of his journey, he had his hands on the wheel for just 25 seconds; he was relying on the car's software to control the vehicle for the rest of the time. When a truck pulled out into the road, neither Brown nor the software spotted it, and the car crashed into the truck.

Joshua Brown became the first person in the history of the world to die in a self-driving car accident.

Welcome to the future.

EPILOGUE

FUCKING UP THE FUTURE

In April 2018, a deal was announced to reopen a previously closed coal-fired power plant in Australia. This was unusual for obvious reasons—as the world tries to slowly move away from climate-change-causing fossil fuels, reopening a coal-burning plant seems a strange move—but it was even more unusual because of the main impetus for it being reopened. It was to provide cheap power to a company mining cryptocurrency.

Bitcoin is the most widely known of the cryptocurrencies, but there's an ever-expanding ecosystem of the things as companies launch new ones at a seemingly exponential rate, hoping to cash in on the mad scramble for digital money. These currencies aren't "mined" in the way that, say, gold is. They're just bits of computer code, most of them based on something called blockchain technology, where each virtual coin is not just an item of symbolic value but also a ledger of its own transaction history. The computational power needed to create them in the first place, and to process their increasingly complicated transaction logs, is significant—and, as such, sucks up electricity at a crazy rate, both to run the ever-larger data centers devoted to crypto-mining and to cool them down as they overheat.

Cryptocurrencies don't have any intrinsic value, and by design most don't have any kind of central authority to regulate and control their flow. The only limiting factor is the cost of the computing you need to do to create and exchange them. But the belief among some people that they're the currency of the future has led to many cryptocurrencies surging in value, as everybody agrees that they're worth something—or, at the very least, that there'll be another sucker along in a minute who thinks they're worth more than you do, until all of a sudden there isn't. So their value has become wildly volatile, depending entirely on the mood of the market. It's a classic financial mania, bubbles forming and bursting over and over again, as everybody tries to not be the one left holding a suddenly worthless parcel when the music stops.

But like most manias, it has real-world effects. It's not just Australia reopening a power plant: in the rural west of America, 170 years after the gold rushes first brought people flooding west, tempted by the prospect of overnight riches, there's a new gold rush happening. Lured by cheap power and cheap rent and space to build, cryptocurrency firms are investing hundreds of millions creating huge, power-hungry crypto-mines in small country towns across Washington, Montana, Nevada and more. Residents of one town where these twenty-first-century prospectors have moved in complain that the around-the-clock roar of the servers is keeping them awake, affecting their health and driving away local wildlife.

By the end of 2018, one estimate predicts that Bitcoin mining alone will use as much energy as the entire country of Austria.

This book has been about the failures and mistakes we've made in the past. But what about the mistakes we're making right now, and the ones we'll make in years to come? What shape might the fuck-ups of the future take?

Making predictions is, as we've noted, a sure-fire way to make yourself look stupid to historians further down the line. Maybe the decades and centuries ahead will see humanity commit a whole series of completely original, novel mistakes; maybe we'll find a way to stop making mistakes at all. But if you were of a mind to put money on it, a sensible bet would be that we'll probably carry on making the exact same mistakes as we have in the past.

Let's start with the obvious, then.

Of all the stuff we've just casually dumped into the environment on the grounds that, eh, it'll probably be fine, it's the carbon we've been merrily burning up since the Industrial Revolution kicked off that's going to really spoil everybody's good times.

That man-made climate change is real, and potentially an existential threat to many communities around the world and many aspects of civilization, is so well established as a scientific fact by this point that it seems kind of dull to run over the evidence again. We're way beyond the point where this might all be another polywater or N-ray-type situation that everybody's going to be embarrassed about in a few years' time. And yet apparently there are still plenty of people who have enough reasons to deny it—financial, political, the sheer bloody-minded joy of being a contrarian knobber—that we keep getting dragged back to the "debate about whether it's real" stage every time it looks like we might make some progress on the "actually do something about it" stage. It's pretty much the playbook the manufacturers of leaded gasoline used back in the day: you don't need to disprove something, you just need to be able to claim the jury's out long enough to keep raking in that sweet, sweet profit.

And so we're doing the collective version of "lalalalala can't hear you," when instead we should probably be running around in a panic like our house was on fire, which...it sort of is. Seventeen of the hottest eighteen years on record have occurred since the year 2000. For the first time in our geological epoch, the level of carbon dioxide in the atmosphere crossed the threshold of 410 parts per million in April 2018. The last time it was this high was during the mid-Pleistocene warm period, around 3.2 million years ago—right when Lucy was falling out of her tree. In case you're thinking, oh well, if it's been that high before, it can't be too bad, at that time sea levels were over 60 feet higher than they are today.

Oh, and climate change isn't the only thing that carbon dioxide is doing. In fact, one of the things keeping a check on the amount of CO_2 in the atmosphere is that the oceans absorb some of it. Good news, hey? Turns out not. Ocean water, much like your boyfriend, is fairly basic—which is to say, it leans very slightly more toward the alkaline than the acid. But absorbing all that CO_2 turns it more acidic, and the more acidic the ocean is, the greater the knock-on effect on marine life, from tiny mollusks to big fish.

Oh, also that gets worse if it happens in combination with the oceans warming up. Which they are. If you want an example of just how bad things are getting in the water, the Great Barrier Reef—one of the *actual wonders* of the *actual natural world*—is dying at an alarming rate, with two years in a row of massive "bleaching" events killing corals across large sections of the reef.

Guys... I think we might have fucked this up a bit.

Of course, that's far from the only doom we've been energetically and determinedly setting ourselves up for. We've given ourselves options here, people. For example: in May 2018, it was reported that scientists had detected a sharp rise in chlorofluorocarbon emissions. Somewhere in the world, likely in Asia, someone's started manufacturing Thomas Midgley's supposedly banned invention again. It could set the ozone layer's recovery back by a decade. Good work on the "learning from our mistakes" front, guys.

Or take antimicrobial resistance. Antibiotics and other antimicrobial medicines were one of the greatest steps forward of the twentieth century, saving countless lives. But, like the people of Easter Island cutting down their trees, we used them too much, too often. The thing is, every time you use an antibiotic, you're increasing the chances that one of those microbes will be resistant to it—and then you're just killing off their competition. It's accelerated evolution, as our actions breed new strains of antibiotic-resistant superbugs that have the potential to bring all the bad old diseases of history flooding back (and it doesn't even need the tundra to melt to do it).

As a result, the world is rapidly running out of effective antibiotics—and part of the problem is that antibiotics simply aren't profitable enough for drug companies to invest sufficient resources into creating new ones. One estimate suggests that already 700,000 people every year die from antimicrobial-resistant diseases.

Or maybe our downfall will come because we keep outsourcing our decisions to computer algorithms, in the hope that this somehow makes them better and wiser, and that it won't be our fault when they go wrong. The algorithms that control self-driving cars are just one example: elsewhere algorithms are deciding what stocks to buy and sell, what news we see on our social media, and how likely it is that someone convicted of a crime will reoffend. We like to think that these will be more rational than humans; in reality, they're just as likely to amplify all the biases and faulty assumptions we feed into them.

The worry about outsourcing our decisions to computers doesn't stop there, as research on artificial intelligence progresses apace. The fear here is that if we do manage to create an AI that's far smarter and more capable than humans, we might be mistaken in thinking it'll be on our side. It might be able to manipulate us to its own ends, it may see us as a threat and destroy us or it may simply fail to recognize that humans are important, and we'll end up being little more than fodder for its goal of creating as many paper clips as possible (or whatever other task we've set it). The prospect that we might Frankenstein ourselves into oblivion may seem remote, but a worryingly large number of supposedly smart people appear to be taking the prospect quite seriously.

Or maybe we'll just blow ourselves up in a nuclear war before any of this happens.

Or perhaps the fuck-up won't be as dramatic as that. Maybe we'll just quietly doom ourselves to a crappy future through our own laziness. Ever since we slipped the surly bonds of earth and entered the space age, our approach to stuff we don't need anymore in space has been pretty much the same as our approach to all the other rubbish we create: we just throw it away. Space is very big, after all, so how much would it matter?

That's where Kessler syndrome comes in. This was predicted by NASA scientist Donald Kessler way back in 1978, and yet it hasn't stopped us chucking stuff away in space. The problem is that when you dump things in orbit, they don't really go anywhere. It's not like throwing a potato chip bag out of a car window, to be immediately forgotten about—space rubbish stays orbiting at roughly the same speed and on the same trajectory as the thing it was thrown out of. And sometimes it will collide with other bits of junk.

The trouble with that is that, because of the speed objects in orbit are moving at, a collision becomes incredibly destructive. A single collision with the smallest piece of material can be catastrophic, destroying satellites or space stations. And those deadly collisions produce, that's right, thousands upon thousands more pieces of space junk, all of which can cause more collisions. This is what Donald Kessler predicted: that eventually space will become so crowded that this process will reach a tipping point, where each collision creates more and more collisions, until our planet is surrounded by an all-enveloping cloud of high-velocity garbage missiles. The result of this: satellites become useless, and launching into space becomes a deadly risk. We could become, effectively, earthbound.

In some ways, it feels like it would be a weirdly poetic end to the journey that Lucy failed to begin all those millions of years ago. All that exploration, all that progress, all those dreams and grand notions, and that's where we end up: trapped on our planet by a prison we've made from our own trash.

Whatever our future holds, whatever baffling changes come along in the next year, the next decade and the next century, it seems likely that we'll keep on doing basically the same things. We will blame other people for our woes, and construct elaborate fantasy worlds so that we don't have to think about our sins. We will turn to populist leaders in the aftermath of economic crises. We will scramble for money. We will succumb to groupthink and manias and confirmation bias. We will tell ourselves that our plans are very good plans and that nothing can possibly go wrong.

Or...maybe we won't? Maybe this is the moment that we change, and start learning from our history. Maybe all this is just being pessimistic, and no matter how dumb and depressing the world today may sometimes seem, in actuality humanity is getting wiser and more enlightened, and we are lucky enough to live at the dawn of a new age of not fucking up. Maybe we really do have the capacity to be better.

One day, perhaps, we'll climb up a tree and not fall out.

★ ★ ★ ★ ★

ACKNOWLEDGMENTS

I couldn't have written this book without the help of a lot of people. My thanks in the first place must go to my agent Antony Topping, without whom I would very literally never have written it. Alex Clarke, Kate Stephenson, Ella Gordon, Becky Hunter, Robert Chilver and the whole team at Headline were a delight to work with, and I'm very sorry about the deadlines. I'd also like to thank Will Moy and the wonderful people of Full Fact for, among other things, waiting too long.

My family—my parents, Don and Colette, and my actual proper historian brother, Ben—were supportive throughout. Hannah Jewell provided funny history book inspiration, insight and a shared understanding of ghosts. Kate Arkless-Gray offered canny advice, a sympathetic ear and also, crucially, a really cracking housesitting opportunity. Maha Atal and Chris Applegate offered stimulating discussion and numerous suggestions; Nicky Reeves likewise. I also have to thank the historians of Twitter for being reliably great and supportive—in particular Greg Jenner (who I loosely paraphrase on page 9) and Fern Riddell; please buy their books, too. Just going to keep adding people now to make it look like I have lots of friends. Damian and Holly Kahya, James Ball, Rose Buchanan, Amna Saleem and many others provided wise words and pints. Bumping into Kelly Oakes repeatedly while in the final stages of writing was exactly the motivation I needed to keep going. I'd also like to thank Tom Chivers for the lunch we never had, sorry about that. The band CHVRCHES released a strong album during the writing process; I include them in this section merely in the hope that someone will distractedly skim it for names but not really take in the context, and as such will think my life is a lot more glamorous than it seems. On that basis, my thanks also to Beyoncé, Cate Blanchett and Ghost David Bowie.

It goes without saying that any failures in this book are mine alone, and do not reflect on any of them. Except for Ghost David Bowie.

FURTHER READING

There are a number of books that I want to acknowledge that I leaned on particularly heavily for certain sections of the book. (Some are already referenced in the text.) All of them are well worth reading for a deeper dive into some of the issues and events that the space in this book only allowed me to touch on.

Daniel Kahneman's *Thinking, Fast and Slow* is mentioned in the section on cognitive weirdnesses, and underpins a lot of our understanding of how our minds operate. Meanwhile Robert E. Bartholomew's *A Colorful History of Popular Delusions* is a great read on manias, crazes, fads and panics.

Jared Diamond's *Collapse* is also mentioned in the text, and heavily informed the section on Easter Island (moreover his influence is clear throughout that whole section).

Volker Ullrich's *Hitler: Volume I: Ascent 1889–1939* was a source for much of the Hitler material (and fans of elegant literary subtweets will also recognize the central conceit of Michiko Kakutani's spectacular review of the same book).

Another one that's referenced a couple of times in the text is Douglas Watt's *The Price of Scotland: Darien, Union and the Wealth of Nations*, which is an insightful and meticulous unpicking of William Paterson's folly.

Frank McLynn's *Genghis Khan: The Man Who Conquered the World* and Jack Weatherford's *Genghis Khan and the Making of the Modern World* were important for the Khwarezm section.

I also want to shout out a pair of books that trod similar ground before this one: Bill Fawcett's *100 Mistakes That Changed History: Backfires and Blunders That Collapsed Empires, Crashed Economies and Altered the Course of Our World* and Karl Shaw's *The Mammoth Book of Losers*, both of which were delightful reading and introduced me to several excellent fuck-ups I wasn't previously aware of.

PICTURE CREDITS

Getty Images: p. 58 (PhotoQuest); p. 63 (Bettmann); p. 65 (© Jim Xu); p. 79 (Bettmann); p. 81 (© Walter Astrada/AFP); p. 107 (© Robert Preston/Age Fotostock); p. 151 (Keystone); p. 179 (Hulton Archive); p. 180 (Historical Map Works LLC and Osher Map Library); p. 242 (Corbis Historical); p. 282 (© Nicholas Kamm/AFP)

Alamy: p. 102 (SOTOK2011); p. 105 (World History Archive); p. 210 (Granger Historical Picture Archive); p. 266 (Granger Historical Picture Archive); p. 267 (IanDagnall Computing)

Bridgeman Images: p. 213 (Bibliothèque nationale de France, Paris [MS Pers.113.f.49]); p. 235 (Académie des sciences, Paris/ Archives Charmet)

ABOUT THE AUTHOR

Tom Phillips is a journalist and humor writer based in London. He was the editorial director of BuzzFeed UK, where he divided his time between very serious reporting on important issues, and making jokes.

Over his career Tom has been a member of a very briefly acclaimed comedy group, worked in television and in Parliament, and once launched an unsuccessful newspaper.

He studied Archaeology and Anthropology and the History and Philosophy of Science at Cambridge, and is pleasantly surprised to have written a book that actually makes use of them.

INDEX

Page numbers in *italics* indicate illustrations.